INVASIVE SPECIES

Invasive Aquatic and Wetland Animals

INVASIVE SPECIES

Invasive Aquatic and Wetland Animals
Invasive Aquatic and Wetland Plants
Invasive Microbes
Invasive Terrestrial Animals
Invasive Terrestrial Plants

INVASIVE SPECIES

Invasive Aquatic and Wetland Animals

Suellen May

CHELSEA HOUSE PUBLISHERS
An imprint of Infobase Publishing

Invasive Aquatic and Wetland Animals

Chelsea House
An imprint of Infobase Publishing
132 West 31st Street
New York NY 10001

Library of Congress Cataloging-in-Publication Data

May, Suellen.
Invasive aquatic and wetland animals / Suellen May.
p. cm. -- (Invasive species)
Includes bibliographical references and index.
ISBN 0-7910-9129-5 (hardcover)
1. Aquatic animals--Juvenile literature. 2. Introduced animals--Juvenile literature. I. Title. II. Series: May, Suellen. Invasive species.
QL122.2.M376 2006
581.7'6--dc22
2006011032

Chelsea House books are available at special discounts when purchased in bulk quantities for businesses, associations, institutions, or sales promotions. Please call our Special Sales Department in New York at (212) 967-8800 or (800) 322-8755.

You can find Chelsea House on the World Wide Web at http://www.chelseahouse.com

Text design by James Scotto-Lavino

Cover design by Takeshi Takahashi

Printed in the United States of America

Bang EJB 10 9 8 7 6 5 4 3 2 1

This book is printed on acid-free paper.

TABLE OF CONTENTS

The Natural World of Aquatic and Wetland Animals

1

The origin of life is the sea. Species that existed long before dinosaurs, such as the horseshoe crab, are still present and swimming, slithering and thriving as living fossils. Given the relatively short time life has existed on land, it is no surprise that the majority of life exists in the seas. In fact, nine-tenths of the Earth's living things live in the water.[1] Aquatic systems support life on land. The very reason life exists on Earth is because it is the only planet we know of with liquid water on its surface.[2]

Aquatic systems aid human activities. Most of the major U.S. cities are near a substantial river or body of water (Figure 1.1). Water provides a means of commerce, recreation, food, medicines, and many other resources impossible to list here. The life contained within these aquatic communities keeps the activities going. The oceans store about 1,000 times more heat than does the atmosphere.[3]

WHAT'S IN A NAME?

Humans like to put things into categories. Putting living organisms into increasingly narrow (or increasingly larger, depending on how you look at it) categories helps us to decide how they are related and which attributes an organism has based on the category it is in. This ordering is referred to as taxonomy. The relationship between the categories is a hierarchy, and the classification system of living things is referred to as a

Figure 1.1 Waterways provide recreation, commerce, food, and vital habitat for many plants and animals. Streams and their banks are usually more biologically rich in both numbers and species diversity than drier habitats.

hierarchical taxonomy. Without it, we would be left with 9 or 10 million species of animals on Earth to name without the help of a system to determine which species are more closely related.

Remember this saying to help recall the basic levels of hierarchy in our classification system for organisms: King Phillip Came Over From Germany Sad. The first letter of each word in this saying correlates with the following: kingdom, phylum, class, order, family, genus, species. These categories will be used throughout the book to indicate how closely related different animals are.

The last two groups in the hierarchy, genus and species, are the ones used to give a creature its scientific name, which is always in Latin. Animals usually have a common and a scientific name. The sea lamprey is a common name for the animals in the genus *Petromyzon* and the species name is *marinus*; therefore, the scientific name of the sea lamprey is *Petromyzon marinus*.

The scientific name often appears in parentheses after the common name.

If we go to the top of the hierarchy, the sea lamprey is in the animal kingdom because of attributes such as being multicellular and needing to consume other organisms for survival, unlike plants, which can make their own food. These are the criteria that scientists use to classify animals. Sea lampreys represent one of the oldest groups of vertebrates, having diverged from other vertebrates millions of years ago. Their vertebrae may be considered primitive, but the presence of those vertebrae puts them in the phylum Chordata, which includes modern fishes, and the subphylum Vertebrata, which includes mammals.

It may seem easier to refer to an animal by its common name. A common name usually reflects a physical attribute or region that the animal is from. Most natural resource professionals do in fact use common names to refer to animals. The problem with relying solely on a common name is that most species have more than one. The Asian clam is also called the Chinese clam, but it has one scientific name: *Potamocorbula amurensis.* Scientists understand from this name that it is a mollusk, among other things. The Asian clam, as it is commonly referred to in the United States, is also called the Amur river clam, Amur river corbula, Asian bivalve, brackish-water corbula, and marine clam. In Germany, the common name is Nordpazifik-Venusmuschel, and in Japan, they say Numakodaki. Using a scientific name enables scientists from all regions of the world to know without ambiguity which organism is being referred to and the accompanying attributes based on the classification.

All life is grouped into categories based on similarities. Until very recently, the broadest category that life was placed into was kingdoms. Since 1969, the scientific community has recognized five kingdoms: Monera, Protista, Fungi, Plantae, and Animalia. Prior to 1969, there were only two kingdoms: Plant and Animal.

The kingdom **Monera** includes bacteria and blue-green algae. The kingdom **Protista** includes mostly unicellular organisms such as algae and amoebas. The kingdom **Fungi** includes mushrooms, molds, and mildews. The kingdom **Plantae** includes mosses, ferns, and flowering plants. The kingdom **Animalia** is broken down into **coelenterates**, which are organisms with a hollow body cavity such as jellyfish; **annelids**, which are worms with segmented bodies; **arthropods**, which have exoskeletons and jointed legs, such as grasshoppers, lobsters, and spiders; and **chordates**, which have a nerve cord and usually a backbone, such as humans and other mammals, birds, reptiles, amphibians, and fish.

The five-kingdom system is being revised to expand the number of kingdoms and put them under another category of "superkingdoms" referred to as domains. The domains are Bacteria, Archaea, and Eukarya. The reason for creating an additional taxonomic category is that scientists have discovered that unicellular organisms without a nucleus (organisms in the Bacteria and Archaea domains) are far more varied than they once thought. Multicellar organisms with a nucleus (organisms in the Eukarya domain) are plants, animals, fungi, and protists.

The number of aquatic species is far too large to list in an introduction. However, remarkable attributes specific to aquatic and wetland animals are easily pinpointed. Below is a brief introduction to the features of the aquatic and wetland animals and how we, along with other animals, rely on them.

RAISING CONCH

In the Caribbean islands of Turks and Caicos, the world's only conch farm can be found in the scrub-like vegetation near a bay. The owner of the farm, trained as a marine biologist, was shipwrecked back in the 1980s while he was boating in the Caribbean. He was enchanted by the aqua seas and made this British territory his home. Then he started a conch farm.

Visitors to the conch farm can still see the broken boat, the wood sun-bleached and cracked.

Conchs are mollusks and are in the phylum Mollusca, along with the octopus. The shell is the true indicator that this animal is a mollusk. Conchs are a favorite food for lobsters, stingrays, crabs, octopuses, and turtles. In the wild, only one small conch out of 500,000 eggs in a single egg mass might make it to adulthood. The rest get eaten.

The conch farmed at the Caicos farm is the Caribbean queen conch (*Strombus gigas*) and is a beautiful, pink-lipped shellfish that is the national dish in Turks and Caicos (Figure 1.2). The seafood that is contained within the shell is high in protein and is a common appetizer served deep-fried as "conch fritters" at many of the island's resorts.

Figure 1.2 This Caribbean queen conch is a mollusk that is easily identifiable by its pink-lipped shell. Like all mollusks, the Caribbean queen conch begins life as an egg. Only one in 500,000 conch eggs survives to adulthood; the rest are eaten by a variety of aquatic animals.

Like all mollusks, conchs start as minute eggs. Egg masses contain 500,000 eggs each and are gathered by divers from the egg farm and brought to the hatchery in late spring. This farm's hatchery resembles a greenhouse. The hatchery can produce 6 million larvae annually from the eggs. The larvae become baby conchs, or **veligers**, after a short period of time. At the Caicos farm, 500,000 veligers are produced each month.

Conchs grow very slowly. Juvenile conchs mature in the post-larval facilities, where they grow from .15 inch to .75 inch (4 millimeters to just 2 centimeters) in length. The young conchs are fed tiny pellets of food during this growth stage. From the post-larval facilities, the .75-inch-long (2-centimeter-long) conchs graduate to the onshore nursery pond and live there for the next year, where they will grow to 2.25 inches (6 centimeters). At 2.25 inches (6 centimeters), the conchs are ready to get their first taste of the sea by being introduced to the subsea pasture, which is a fenced portion of the Caicos Bank adjacent to the conch farm. A conch has reached adulthood when it has grown its distinctive, beautiful pink lip. Then it will be harvested and served as a high-protein delicacy.

Mollusks, just like many other seafaring species, are on the decline in many parts of the world because of overfishing. In the waters of the Caicos Bank alone, 4 million conchs are supplied each year, but that number is declining. Luckily, regulations exist to stop the decline of species before it is too late. The conch has been listed as an endangered commercial species by the Convention on International Trade on Endangered Species (CITES). This international organization, set up in 1975, works to control the exploitation of endangered species.

MERMAIDS AND SEA MONSTERS

Extending 200 feet (61 meters) deep within a sinkhole on a tiny island in the Caribbean, is an aquatic ecosystem that is so mysterious that the island's inhabitants have spun tales of a mermaid (Figure 1.3). If a mermaid were to exist here, she would have to endure water temperatures hotter than a Jacuzzi, overcome a very limited food supply, be able to see in the dark, and withstand a 50-foot (15-meter) layer of foul-tasting sulfurous water. Perhaps unappealing as a home for a mermaid, this unique environment is nevertheless a place where some creatures thrive. One such creature is the beautiful sea monster.

The beautiful sea monster (*Kaloketos pilosus*) is found nowhere else on Earth. Fossil records of a close relative prove that it existed while dinosaurs roamed the planet. Although it is only 1.15 inches (3 centimeters) long, the beautiful sea monster is a predatory **crustacean** that lives in the darkness and attacks, then kills, creatures much larger than itself. Once it has seized its prey, it uses fang-like limbs to inject a poisonous digestive fluid into the prey. The beautiful sea monster then eats its prey alive.[4]

Cottage Pond is the name of this sinkhole that is part of a deep network of underground caves that extend all the way to the ocean miles away. Sinkholes are formed when the underlying layer of limestone dissolves. In Cottage Pond, freshwater and saltwater mix to become **brackish water**. This meeting of freshwater and saltwater is what causes the rotting layer of sulfurous water. Known as the mixing zone, this region is created by rotting vegetation that sinks through the freshwater layer of the pond but then bounces off the denser saltwater layer like a trampoline.[5] The red soupy water is so acidic that it burns human skin. Although it seems that this type of environment would restrict life, this sinkhole is actually teeming with it. This is the environment that certain species need to live, particularly the beautiful sea monster. How this crustacean copes with living in such an extreme environment has yet to be learned. Scientists do know, however, that this crustacean evolved with its environment. To alter even one

Figure 1.3 Aquatic animals can thrive in surprising locations such as this deep sinkhole. A sinkhole is often the only visible portion of a large underground network of caves and may be home to aquatic species that exist in extreme conditions, many living in the absence of light.

aspect, even the sulfurous mixing zone, is to risk losing a newfound species.

HIBERNATION

It is winter in Minnesota and William Schmid walks through a leafless woodlot. The frozen ground and fallen foliage crunches underfood. Schmid is no average naturalist. He is a professor of physiology at the University of Minnesota.

Schmid has studied frogs and their responses to dehydration. As he walks, he sees a wood frog hidden under the fallen leaves. Schmid picks up the frog and notices that the frog does not do what all frogs do in the hands of an observer: wink. Puzzled, Schmid thinks the frog looks frozen. Knowing what he knows about frogs, Schmid doubts that the frog merely neglected to bury itself deeply enough for winter. Hibernation is what the scientific community accepts as the frog's method of surviving the freezing temperatures of winter. Schmid believes that this frog is somehow able to withstand freezing as a means of getting through the winter. Schmid's leisurely stroll, which occurred in the early 1980s, is the first documented field observation of freezing-tolerance of any vertebrate.

When a frog freezes, there is extensive ice formation. Ice is dangerous because it can cut like a knife, slashing membranes, puncturing cell components, and breaking cells.[6] Frogs that spend the winter frozen only form ice in the body cavity and not in the cells. The frogs prevent the ice from forming in their cells by producing a glucose-like chemical that acts as antifreeze. The amazing aspect of this is that the frogs do not make this substance in anticipation of winter. Instead they make this chemical in a rush when they feel their body start to freeze. The glucose-like substance helps to sustain the frog until spring's thaw, at which time it will move to a watery breeding site to mate.

Frogs are amphibians, which have the distinction of living both on land and in the water. Amphibia translates to "both lives." Amphibians are cold-blooded vertebrates that undergo a metamorphosis from an aquatic, water-breathing, limbless larva (the tadpole stage) to a terrestrial or partly terrestrial air-breathing, four-legged adult.[7]

MARINE MAMMALS

Marine mammals spend most or all of their time in the water.[8] Life evolved from the sea and the earliest species were aquatic. Mammals, however, evolved on land. Marine mammals, including whales, sea otters, manatees, seals, sea lions, and dolphins, are mammals that moved to the sea many millions of years ago (Figure 1.4). In the process of adapting to aquatic life, they lost rear limbs that were possessed by their land-dwelling ancestors. Their bodies resemble fish, and they still retain many of the characteristics, such as hair, of their land-dwelling ancestors—so much so that many people often forget that a whale is not a fish but a mammal.

Aquatic mammals have the dilemma of breathing air and yet living in water. Certain adaptations are required for these mammals to survive in their aquatic world. Penguins and crocodiles have a movable flap in their throat that closes when the nostrils take in air.[9] Aquatic mammals have the unique ability to hold their breath, which very few creatures can do. This is a valuable adaptation for mammals such as nutria (*Myocastor coypus*) that swim very fast and far underwater to elude capture.

Mammals also give birth to live young and nurse them, which forges a more nurturing relationship with their young. "Bonding" chemicals, such as oxytocin, are produced during nursing. Bonding ensures that the mother will be more protective of her young, ensuring a better rate of survival. Fur seals in the

Figure 1.4 Manatees belong to the unique evolutionary group known as marine mammals. Life began in the sea and then evolved to the land, including mammals. Marine mammals returned to the water, retaining some features common to all mammals such as hair, mammary glands, and the characteristic of giving birth to live young.

Galápagos Islands off the west coast of Ecuador are plentiful. Visitors can watch the interactions of a mother seal and her pup. When the mother goes out to sea to fish for her baby, the pup whines and goes from one seal to another crying for its mother.

SKELETONS: INNIES, OUTIES, AND SOFT-BELLIES

The earliest life-forms were not only aquatic but were microscopic and soft-bodied. Soft-bodied means that there is no skeletal system or vertebrae. These simple life-forms never ventured on land and had the natural buoyancy of water to support their tiny structures.

As life-forms became more varied, the advantages of a skeleton became apparent. Skeletons evolved to support the organism's larger or terrestrial structure. Early vertebrates had a primitive rod called a **notochord**, which was the precursor to the spinal column. As the vertebrates evolved, features such as a digestive tract and a vascular system became associated with having vertebrae.

The animal kingdom includes vertebrates and invertebrates. **Vertebrates** are animals with backbones and are in the subphylum Vertebrata. Vertebrates include humans, amphibians, reptiles, and fish. Of all the known animal species, vertebrates make up only about 2%. **Invertebrates** are animals without a backbone that either have a jointed exoskeleton or are soft-bodied. A sponge is a soft-bodied invertebrate; crustaceans with their hard shells are invertebrates as well. The skeleton of vertebrates is buried under skin and muscle; the skeleton of invertebrates is only covered by eyes, antennae, legs, part of the digestive tract, and respiratory surfaces. Vertebrates wear their skeletons on the inside whereas invertebrates with exoskeletons wear them on the outside.

INTERACTIONS IN THE AQUATIC WORLD

Aquatic animals have a variety of interactions with individuals of their own species, other species, and their environment. Some aquatic animals provide some benefit to another organism while getting something out of it themselves, which is called **mutualism. Parasitism**, on the other hand, is a relationship where one gains and the other loses. The study of these relationships is the science of ecology.

Commensalism is a relationship between two species where one species benefits and the other is unaffected. It is difficult to prove that a relationship is true commensalism, because often the second species will seem unaffected but further investigation shows that some effect on the second species actually does occur. Algae growing on a tree is one example of commensalism: The algae has a place to grow without entering or affecting the tree.

In the aquatic world, the clownfish (*Amphiprion percula*) has a close relationship with sea anemones (Figure 1.5). The sea anemone is a marine animal that consists of a hollow cylinder surrounded by a crown of tentacles. The tentacles have tiny structures on them that act like miniature harpoons. These structures contain a poison that can kill or paralyze small fish and other small aquatic animals. The sea anemone creates this poison to protect itself; however, it does benefit the clownfish as well. The clownfish has a protective substance in the mucus of its skin and is unaffected by the sting of the anemone's tentacles.

Figure 1.5 This clownfish finds protection living in the underwater maze of sea anemone tentacles. A protective mucous covering on the clownfish protects it from the poisonous sting of the sea anemone's tentacles. A relationship where one species benefits and the other species is unaffected is known as commensalism.

The clownfish lives in the tentacles and is safe from predators. The clownfish benefits with protection from predators and the sea anemone does not suffer by having the clownfish reside in its tentacles. This is an example of commensalism because one species benefits and the other species is unaffected.

Mutualism is a relationship where both species benefit. It is also referred to as a symbiotic relationship. Species may exchange food or provide shelter, but both may be able to live independently. Although coral may look like beautiful, underwater rock, it is actually a living, aquatic animal (Figure 1.6). Coral lives a symbiotic life. Within the coral lives algae that gives off oxygen that the coral needs to live. The coral provides carbon dioxide and other substances to the algae. That is why you often see coral growing in shallow waters; the algae need

Figure 1.6 Coral is an aquatic animal that lives in the shallow parts of oceans. Coral has a symbiotic relationship with algae. The algae growing on the coral polyps provides oxygen to the coral. The coral provides carbon dioxide to the algae and affords it a place on which to grow.

Nutrient-Rich Waters: Too Much of a Good Thing?

We have all heard the expression, "everything in moderation." This expression also applies to the natural world. When an overabundance of nutrients is present in water bodies, algae and aquatic weeds will grow and consume oxygen in the water. Oxygen is an important component in all bodies of water for biotic communities, including fish. Biotic communities consist of all of the living organisms in a specified area. All these organisms have a demand for oxygen. If algae and aquatic weeds grow excessively, there will be little oxygen remaining for other species.

Fertilizers from farms and lawns can contribute to these nutrient-rich waters, known as eurotrophication. Eutrophication is the process where a lake, stream, pond, or any other body of water becomes high in nutrients, thereby decreasing the amount of dissolved oxygen available for aquatic organisms.

Fertilizers are high in nutrients and although they are valuable when the appropriate amounts are applied to desirable vegetation or crops, fertilizers in water are pollutants. Failing septic tanks can also create eutrophication because the waste products are high in organic matter. The more polluted a section of water, the more microorganisms and aquatic life compete for the oxygen in the water. Therefore high levels of pollution, even in the form of nutrients, can cause fish to suffocate from lack of oxygen. Other man-made activities that lead to eutrophication are factory and cannery effluents, domestic waste discharges, feedlot runoff, and construction site erosion.

enough sunlight to photosynthesize, just like other plants. Crabs will use sponges and seaweed as a way to camouflage themselves from predators. They pick them up with their claws and place them on their backs. In return, the sponges and seaweed acquire food on the backs of the crabs.

Animal species interact with one another, but so do environments. Terrestrial and aquatic environments interact and have a significant impact on aquatic and wetland animals. Contaminants that enter the aquatic system from land are one way that a terrestrial environment affects the aquatic system and its residents. Agricultural and urban runoff is one way that pollutants can directly enter a body of water. If a farmer applies too much fertilizer or too many pesticides, a heavy rain can wash these chemicals into an adjacent water system. Once pollutants have entered the body of water, aquatic animals, particularly filter feeders, incorporate them into their tissues. Filter feeders are animals that get their food by picking up tiny organic material from the water they are living in. They are at the bottom of the food chain. A typical food chain in an aquatic system might start with plankton that gets eaten by a small fish that in turn gets eaten by a predatory fish and then finally is caught and eaten by a human. Plankton is the microscopic aquatic organisms that drift with the waves and currents of the sea.

Pollutants exist in the water and are concentrated in the living tissues of animals. The amount of contaminants in each organism increases the higher the organism is in the food chain. **Biomagnification** is the exponential increase in chemical pollutants from the water through living organisms. Chemicals also can enter an aquatic system by way of atmospheric deposition, industrial outfalls, and migration through groundwater.

The Threat of Invasive Aquatic and Wetland Animals

2

Roger Lang gave up his fast-paced life in California to climb Montana waterfalls in search of westslope cutthroat trouts (*Oncorhynchus clarki lewisi*). Not just any westslope cutthroats will suffice for Lang: He is in search of genetically pure westslope cutthroats. In the early nineteenth century, these trout were plentiful in the discovery days of Lewis and Clark's expedition west (just notice the Latin name given to the trout). Westslope cutthroats are on the decline because of the more aggressive, non-native brown and rainbow trout (Figure 2.1). The non-native trout were competing with the native trout for food and habitat but were also mating with the native trout, making the pure version of the native trout even more rare.

When Lang realized how troubled the native trout were, he sold his software company for $300 million and bought an 18,000-acre ranch in Montana to protect the native trout. By searching Montana streams, he was able to find what he thought were the distinctive trout, so named for the red slash under the fish's jawline. He had genetic testing done to make sure that that they were in fact the native trout and not a hybrid. Trout that were 100 percent pure became part of Lang's captive breeding program. Lang now has hundreds of breeding pairs and 4,000 fingerlings.[10]

Figure 2.1 Westslope cutthroat trouts were once plentiful in the early exploration days of Lewis and Clark. This species of fish is now on the decline because of the introduction of the invasive brown trout and rainbow trout that are not native to the United States.

NATIVE AQUATIC AND WETLAND ANIMALS

Native aquatic and wetland animals spend all or a majority of their life in an aquatic system and have evolved naturally in their environment. They have predators and are part of the food chain. Coral, an aquatic animal, consumes zooplankton (microscopic aquatic animals) that float in the sea. Some scientists use the term *native* to describe an animal living in its country of origin; others use the term more strictly to mean living within a specific ecosystem.

The definition of a native animal is complicated because animals can move from one country to another through migration and still be in their native environment. For thousands of years, whales have journeyed from the frigid Atlantic to the Caribbean sunshine in the winter. They are considered visitors to the Caribbean tourists who see them while on whale-watching trips. In the late winter to early spring, the whales head back to the Atlantic with their young calves at their sides. Although it would be difficult to pick an exact aquatic system that the whales are native to, we do know that they are where they are supposed to be, ecologically speaking. This migratory pattern has been carved out over the years, and it keeps the whale population alive but not growing too far out of control.

INVASIVE AQUATIC AND WETLAND ANIMALS

Invasive aquatic and wetland animals are animals that have moved into an area through purposeful or inadvertent human means and reproduced so aggressively that species that are native to that region are being negatively affected. Invasive aquatic and wetland animals are also referred to more broadly as invasive species. **Invasive species** include animals, plants, and microbes that according to the U.S. Fish and Wildlife Service are infiltrators that enter ecosystems beyond their historic range. Their invasion threatens native ecosystems or commercial, agricultural, or recreational activities dependent on these ecosystems. They may even harm the health of humans. In some cases the native species is completely displaced. Invasive species are usually not native to an area, rather they are usually from another country or region. Other terms used to describe invasive species are *invaders*, *nonnatives*, *exotics*, and *nuisance species*. These terms are used interchangeably.

Invasive species were either accidentally or intentionally introduced almost always without their natural predators. Natural predators keep a population under control. Predators of aquatic animals include disease pathogens and any organism that will eat or parasitize the invasive aquatic animal.

Predators and prey evolve together and occur naturally in the environment. The presence of a predator results in the survival of the fittest of the prey species. Prey will try to avoid being eaten and will therefore develop characteristics to escape, combat, or discourage the predator. The prey may evolve to be faster if the predator is a chasing animal. The prey may also develop better camouflage or an improved sense of smell or sight. Characteristics that do not enable prey to escape a predator will not be improved or refined. Similarly, characteristics that predators have that enable them to catch their prey, like claws, sharp teeth, and keen eyesight, would be developed over time.

Predator-prey relationships are nature's way of keeping species in balance. The predator's population may be larger than usual one season but will decrease because of the limits on food (the prey). Separating the predator from its prey knocks the system out of balance and causes a potentially damaging situation for the environment, a "superspecies" for which there are few controls on the population.

TRAVEL, TRADE, AND TOURISM

Invasive aquatic and wetland animals are not a new problem. Humans have assisted with this invasion for centuries. Many of these invasives entered the United States in the 1800s. In some cases they were intentionally introduced, such as the brown trout, because their introduction was seen as beneficial. Many anglers wanted the brown trout here for sport fishing. A hundred years ago nobody realized the kind of ecological warfare these animals would wage on our native animals.

Today we see the dwindling of native species as a result of these invaders. The introductions may not be new but the problem and its effects are accelerating due to travel, trade, and tourism, according to Lori Williams from the Invasive Species Council.[11]

Whereas oceans and unscalable mountains kept animals from entering terrain where they did not belong, trade and travel have bridged any geographic barrier. A plane can take you from Philadelphia to sub-Saharan Africa. If an animal finds its way into the cargo area of a plane, it could travel thousands of miles in a day. The barriers that nature has instituted can be dismantled in a day.

Trade is another avenue of importation for invasive aquatic and wetland animals. Schools of fish are often inadvertently introduced through ballast water. When a ship is empty of cargo, it sits high in the water and can be more easily overturned by wind or waves. To make travel safer, the ship fills its tank with millions of gallons of water to make it heavier and therefore more stable in transit along coasts and on the open seas. This water is referred to as **ballast water**. Water ends up being collected from one port and then dumped in another.

The problem with ballast water is that species that are native to one port are suddenly the outsiders in another port thousands of miles away. Scientists estimate that as many as 3,000 alien species per day are transported in ships around the world (Figure 2.2). These organisms range from microscopic plants and animals to

Figure 2.2 Trade is an important source of introduction for invasive aquatic and wetland animals. The ballast water contained in a ship to help maintain balance on the high seas can introduce schools of fish and other animals to ports thousands of miles away from their native waters.

crabs, mussels, and schools of fish. Not all of these species survive, but the ones that do survive produce offspring; the population then continues to increase exponentially.

DISCARDED PETS AS INVASIVE SPECIES

The international pet trade is a multibillion dollar business. One popular pet is the Burmese python, native to Burma in Southeast Asia (Figure 2.3). People will pay as little as $20 for a cute hatchling without thinking about how large it will become. In just a few years, this snake can grow to 9 feet (2.70 meters) with a voracious appetite for prey. Most people are not willing to accommodate the snake's appetite and simply toss it in a body of water.

In the 1990s, wildlife officials first noticed the Burmese python in the Everglades. Already an ecosystem that has suffered from development, this new invasive species is having a visible impact. The python is a top-level predator, meaning that almost every other animal is considered prey. The python will eat reptiles, otters, squirrels, woodstorks, and sparrows.

The alligator was once considered the Everglades top predator, but now the python is outcompeting it. Wildlife officials have seen the python's ambitious consumption in play: A 13-foot (3.9-meter) python burst after it tried to swallow a live, 6-foot (1.8-meter) alligator whole. Another problem is that as an introduced species, the python has no natural predator, not even disease, so its population could theoretically grow out of control with no checks in place.

HOW ANIMALS COMPETE

In the 1800s, Charles Darwin was curious about the natural world. As any good scientist would, he observed and took notes. Darwin took an epic voyage from England to South America to discover and observe even more animals.

Darwin believed that different species were not independent creations but rather descended from common ancestors. Present-

Figure 2.3 The international pet trade introduces many invasive animals, including the Burmese python. The python is predator to a range of animals and as a result has disrupted the ecosystem in Florida's Everglades.

day scientists know that mammals evolved from reptiles. Darwin described this idea in his book *The Origin of Species by Natural Selection*. "Descent with modification" was how he described the reason for different species. He explained that in a group of animals, there are slight changes that make them better at competing with other animals or better at living in their environment. These natural changes are referred to as adaptations. **Adaptations** are small changes a species makes over time to better compete in its environment.

Darwin also described the gradual process by which species that are marginally better adapted to their environment reproduce more successfully than those that do not possess a particular minute advantage. This is what he called **natural selection**. According to Darwin's natural selection, the less-well-adapted species gradually disappear and are replaced by those that are more fit. Hence the term survival of the fittest to explain natural selection.[12]

Nature gives better-adapted individuals within a species a better chance to reproduce and therefore increase the number of the fittest of the species. Invasive species thrive not because they have evolved mechanisms over time to make them better equipped to live in their environment but because they have the good fortune of being implanted to an environment where they have no predators to hold them back.

HOW INVASIVE SPECIES MAKE A LIVING

Every aquatic animal has a crucial role in the ocean, lake, wetland, estuary, or pond. Small fish serve as food for predatory fish and consume aquatic plants. By eating aquatic plants, the small fish ensure that the plant population does not grow out of control and make the water too nutrient rich. Some aquatic animals are restricted to very specific habitats, such as the beautiful sea monster described previously. This crustacean would not be able to live in the Delaware Bay or the Colorado River. It is only known to exist in a 200-foot-deep (60-meter-deep)

Minnesota's Malformed Frogs

In August 1995, a group of Minnesota elementary school students caught malformed frogs in a nearby pond. Some of the frogs were missing limbs, some had extra limbs, some were missing digits; there were extra eyes or missing eyes; and some even had skin webbing or malformed jaws.

The scientific community was quickly alerted to this discovery and set out to find a cause. From 1995 through 2000, nearly 25,000 frogs from 195 sites in Minnesota were caught, examined, and released. All the malformed frogs that were found were juveniles, which led scientists to believe that these frogs could not reach adulthood in their condition. This lack of survival may be contributing to the population decline reported for some amphibian species.

Scientists determined that parasites were the cause of some of the malformations. A parasitic flatworm (*Ribeiroia ondatrae*) burrows into tadpoles often where the limbs form, causing limb malformations. The occurrence of this parasite tends to be correlated with higher levels of nutrients. This parasite was found at some of the sites where there were malformations, but not all of them, indicating that there are additional reasons for the malformations.

Scientists have known that certain pesticides can also cause frog malformations. Endocrine disruptors are one kind of pesticide, generally associated with sexual development because they contain estrogen. Scientists found a relationship between the concentrations of estrogenic compounds and the percentages of malformed frogs at some of the sites. Scientists concluded that the malformations are likely a combination of chemical, biological, and physical factors.*

*Sandy Bauers. "Invaders All Around Us." *Philadelphia Inquirer*, October 3, 2004, p.A24.

sinkhole in the tropics. The beautiful sea monster has a narrow ecological niche. An **ecological niche** is the role of an animal in its environment: what it eats, what it does to avoid predators or catch prey, what resources it uses, and how it interacts with other species. Essentially, a niche is how an animal makes a living. An animal with a narrow ecological niche has a limited range of options for food, resources, and habitats. Animals with a narrow ecological niche are also called **specialist species**.

Generalist species are animals with wide ecological niches. Aquatic animals that are generalists have many foods they eat and habitats to live in. An aquatic animal that lives in marine and salty waters has a wide niche. Most invasive species are generalists.

When it comes to survival, evolutionary history has shown that generalists tend to stick around. The horseshoe crab is often referred to as a "living fossil." This was a term coined by Darwin, and although a fossil cannot be living, the name makes the point that the animal's reign reaches back to the days of dinosaurs and trilobites (early marine arthropods).

The horseshoe crab is considered an ecological generalist. This marine animal can withstand polluted waters when most animals have died or left if they could. They can also withstand the salty ocean waters at the Jersey Shore, as well as the brackish waters of an estuary. According to some scientists, this hardiness—that jack-of-all-trades nature that stamps horseshoe crabs so clearly—holds the clue to why they have remained so stable, so evolutionarily nonchangeable, for millions of years.[13] Of course, even the most adaptable animal cannot help but suffer from overfishing, which has recently been the case for the horseshoe crab.

This information is important when a nonnative animal has invaded an ecosystem and its potential impact needs to be evaluated. If the animal is a generalist, this should be a reason for concern. It also helps explain how some exotic species transplanted thousands of miles away thrive and others do not.

Nutria

THE RODENT THAT LIKES TO EAT OUT

3

The Louisiana Department of Wildlife and Fisheries and the National Wetlands Research Center (NWRC) of the U.S. Geological Survey (USGS) are having a cooking contest as a way to help save Louisiana's precious wetlands. Five invited chefs are putting their culinary skills to work in the hopes of winning glass trophies and cash prizes of $100 to $400. It is not an event to raise money or to build camaraderie; rather, it is a creative way to get rid of an unwanted pest—by eating it.

This cooking contest is exclusively for dishes in which nutria is the main ingredient. Nutria are vegetarians, eating only the roots and stems of plants. The meat is high in protein (22.1 percent) and very low in fat (1.5 percent), actually lower in fat than chicken or turkey. The meat is dark and the flavor is easily enhanced with seasoning. Recipes include stuffed nutria hindquarters, nutria chili, and nutria sausage jambalaya. Chefs are surprised at the versatility of this meat.

Around the state, the NWRC has been promoting nutria as a food source as a way to motivate the harvesting of this semiaquatic rodent that harms coastal wetlands. Although some people may not be able to get past the fact that nutria is a rodent, many others in the South have embraced the idea of eating them and have even given nutria dishes the traditional Cajun flair by mixing the meat with sauce piquant or deep-frying it. People have found the nutria dishes to be quite tasty. The Louisiana

Department of Wildlife and Fisheries lists nutria recipes on their Web site, http://www.nutria.com.

BIOLOGY AND BEHAVIOR OF THE NUTRIA

The nutria is listed in the top 100 worst invasive species in the world (Figure 3.1). Nutria are semiaquatic rodents native to South America that were introduced primarily for their fur as early as 1899. Before the nutria's ecological impact was known, sellers promoted these rodents as weed cutters because of their appetite for vegetation.

Nutria are considered large for a semiaquatic animal but are smaller than a beaver. The nutria's tail is round and slightly furry. The clawed digits of the paws are agile and used to dig and excavate roots. This animal can swim long distances underwater with the help of its webbed hind feet and yet is capable of rapid travel on land. Nutria can see well underwater, which helps

Figure 3.1 Nutria are aquatic mammals that can both breathe air and hold their breath underwater. The small, agile body of the nutria enables it to swim fast and far underwater to escape predators.

them escape capture, particularly by people. Their eyes, ears, and nostrils are set high on their heads so they can stay above the waterline when swimming. Valves in their nostrils and mouths can seal out water.

Similar to many invasive aquatic and wetland animals, nutria have high reproductive rates. Females reach sexual maturity between 3 and 9 months, males reach sexual maturity at 4 to 9 months. Breeding occurs year-round. Females give birth to live young and have mammary glands, so they are mammals. With a gestation period of 130 days, a female can have two litters and be pregnant a third time all in one year. The average number of young in a litter is 4 but there can be as many as 13. Newborn nutria are furry with open eyes and are ready to begin feeding on plant roots within hours.

These strict vegetarians eat on land and water by shoving aquatic plants into their mouths with their forepaws. Nutria weigh about the same amount as the average house cat and consume about 25% of their body weight each day. They often construct round platforms in the water from collected plant parts; they use these platforms to feed, birth, rest, and groom.

Nutria also cause extensive damage in levees and dikes by burrowing. Nutria in Louisiana and Texas undermine and breech water-retention levees in flooded fields used to produce rice and crawfish. Nutria burrows can also damage flood-control levees that protect low-lying areas as well as; weaken the foundations of reservoir dams, buildings, and roadbeds. This burrowing also leads to the erosion of banks and streams, lakes, and ditches that are all critical to keeping water in its place.

Introduction and Spread of Nutria

Nutria were introduced to the United States because of their fur. After being introduced to Louisiana's marshes in the 1930s, they escaped their area of confinement and were able to live on their own as a feral population. A previously domesticated animal that is living in the wild is referred to as **feral**. Feral animals are considered invasive because they did not evolve with the environment and tend to be aggressive.

People aided the dispersal of nutria by trapping them and releasing them in marshes in Texas, not that the nutria needed much help, since they were well able to disperse on their own. In 1942, nature helped the nutria. A hurricane came and pushed the nutria inland throughout Texas and Louisiana. At this time, no one knew there could ever be a problem with this animal.

In the late 1940s, nutria were promoted as a way to control water hyacinth, another invasive species. Rapid population growth occurred even though nutria were still being hunted for their pelts. Pelt harvest records provided the main information about the nutria's presence and spread.

It was not until the 1950s that people started to notice damage to marshes in areas where nutria lived. A marsh eaten by nutria doesn't have nearly the vegetation it is supposed to and as a result it erodes. A denuded marsh does not provide a buffer between land and sea or purify water. The damage was not restricted to marshes. The animals damaged other areas including rice and sugarcane fields and levee systems. According to the Louisiana Department of Wildlife, the nutria population soared to 20 million.

A conflict occurred in the 1950s between the fur industry and land managers. Biologists said the nutria were denuding and destroying marshes. The state wanted to promote nutria fur as a natural resource. The fur industry won and nutria were seen as a valuable resource. Hunting nutria for their fur, however, kept the damage to the marshes under control. During the 1960s, 1970s, and 1980s, more than a million nutria were killed for their fur, which helped to keep their populations under control.

In the 1980s, fur became less fashionable and the nutria made a comeback. Biologists observed the damage and aerial photography confirmed severe denuding, particularly to the southeastern marshes. In the 1990s, ground surveys proved that the nutria population was eating away the marshes. Nutria can be found in 40 states and continue to spread.

Down in the Bayou

Nutria live where the water transitions to land: in the wetlands. Wetlands refer to marshes, bogs, swamps, and fens. Despite the name, wetlands are not necessarily wet year-round; some are only wet for 3 weeks out of the year. **Marshes** are wetlands dominated by soft-stemmed vegetation (Figure 3.2). The salt marshes of New Jersey are vast grassy meadows that protect the

Figure 3.2 Marshes are a type of wetland dominated by soft-stemmed vegetation. Marshes invaded by nutria lack an adequate amount of vegetation to support the functioning of the marsh. As a result, the marsh begins to erode and vanish.

mainland from flooding, filter sediments and pollutants from the water, and serve as a nursery for many fish and shellfish.

Swamps are another type of wetland comprised primarily of woody plants. **Bogs** are freshwater wetlands, often formed in old glacial lakes, characterized by spongy peat deposits, evergreen trees and shrubs, and a floor covered by a thick carpet of sphagnum moss. **Fens** are freshwater peat-forming wetlands covered mostly by grasses, sedges, reeds, and wildflowers.

Wetlands are biologically rich with plants and animals. It is surprising how much wildlife food is produced in an acre of wetland. In an experiment, scientists found that a plant important to overwintering birds in South Carolina (*Scirpus robustus*) produced an average of about 300 pounds of seeds per acre per year. Seeds are an important source of nutrition for animals but especially for overwintering birds. Some wetland plants produce as much as 900 pounds of seeds per acre.[14] Birds consume not only the seeds but all parts of grasses, sedges, and rushes including tubers, rhizomes, stems, foliage, and flowers. Wetlands also provide valuable resting ground for migrating birds.

Wetlands also purify water by absorbing excess nutrients, sediment, and other pollutants before they reach rivers, lakes, and other bodies of water. When rivers overflow, wetlands absorb and slow floodwaters. They provide protection from hurricanes.

Louisiana's Wetlands

Before Hurricane Katrina hit the Gulf Coast in 2005, Louisiana's wetlands were a prolific seafood factory, sustaining a commercial fishery that most years generated more than $300 million worth of finfish, shrimp, oysters, crabs, and other delicacies.[15] Louisiana's wetlands were biologically rich and provided a buffer from flooding and hurricanes.

Nutria have been blamed for much of the damage to Louisiana's coastal marshland vegetation (Figure 3.3). Nutria do not just eat the vegetation that holds the soil in place, they eat the very anchor, the plant roots. With their nimble digits, they can destroy a lot of marshland in search of roots. In addition, like most herbivorous animals, they must consume a large amount of food to sustain themselves, amounting to pounds of vegetation per day. With millions of nutria in Louisiana's wetlands, this amounts to tons of precious plants per day.

Nutria are not the only reason Louisiana's marshes and barrier islands are vanishing at a rate of 1 acre every 33 minutes.[16] The problem began when people decided to alter the natural process of the Mississippi River. The Mississippi River is a massive watershed that drains runoff from one-third of the continental United States. A **watershed** is an area drained by a

Figure 3.3 Damage to baldcypress, a long-lived wetland species of tree, by nutria is apparent. Baldcypress is valuable for wildlife food and cover. The nutria's nimble digits and broad vegetarian diet can devastate wetland vegetation.

river. All water flows downhill; as it does, it picks up material, such as sediment. Fast-moving water is able to carry a good bit of sediment, slower-moving water carries less, and perfectly still water carries almost nothing. As water slows, generally at the mouth of a river because it widens and therefore slows, the river deposits this sediment; the resulting fan-shaped deposit is called a **delta** (Figure 3.4). These fresh layers of sediment, which generally occur during spring flooding, are meant to replenish naturally sinking delta soils. Floods are a natural process that benefit natural systems, despite how destructive they can be.

A devastating flood in 1927 raised concern about how these floods could be prevented. Officials decided to build levees along the river and line them with concrete. The result was that Louisiana's wetlands no longer received the replenishing spring flooding that replaced lost sediment. The sediment that the

Figure 3.4 This fan-shaped deposit of sediment is called a delta. Deltas generally occur at the mouth of a river where the water slows and drops its load of sediment. Delta soils must be replenished yearly by flooding.

Mississippi River carried was instead funneled out to the Gulf of Mexico, and thus was lost to the delta and wetlands. With the levees came the construction of ditches that turned the marshes into a giant jigsaw puzzle and increased erosion. To many, this was the beginning of a loss of wetlands in Louisiana so great that no amount of restoration could bring it back to its pre-1920s levels.

The extensive network of oil pipelines has also contributed considerably to Louisiana's wetland loss. Platforms are anchored 18 miles offshore. The construction of the platforms damaged the fragile plants and soil in the wetlands. Also, the oil is pumped out of the muck through a maze of buried pipelines. The slicing of the marshes to accommodate the pipelines is just part of the wetlands loss; the other part is the removal of oil. Locals have compared it to sucking soda through a straw. As you suck, everything goes down. This is not the first time wetlands have vanished from petroleum extraction; grassroots organizations in California continue to fight offshore drilling that has destroyed reefs and coastal wetlands.

You do not have to tell the locals that Louisiana's wetlands are disappearing. They can see it happening in their own backyards. Locals sing about the washing away of the marshes and barrier islands. Elderly locals remember a swamp they could walk through wearing summer dress shoes where now you are knee-high in muck.

Nutria Management

The Louisiana Department of Wildlife and Fisheries is looking at the use of specifically designed baits to control nutria. These baits contain a poison embedded in food. In the past, baits have not been used successfully, so the Louisiana Department of Wildlife and Fisheries solicited the assistance of an organization that could conduct research to determine if using zinc phosphide would control the nutria populations on coastal marshes.

The wildlife managers had a few other questions they hoped the research would answer. They wanted to know what foods the nutria would prefer as bait. They wanted to find out the best substance to use to monitor the nutria's foraging. Since food and the senses are linked for many animals, the wildlife managers wanted to determine what sights, smells, and sounds might attract the nutria to the bait.

Tetracycline, which is an antibiotic, was used in the baits. Tetracycline can have a staining effect on the teeth of mammals (including humans). Research showed that the tetracycline was detectable in the teeth of the dead nutria just two days after the baits were set out. Metallic flakes were also used as "tracers" to tell if a dead nutria had been killed by the bait or just simply died.

Generally, tetracycline is a long-term marker; several weeks elapse between feeding it to animals and tooth collection. Metallic flakes are short-term markers. The flakes are swallowed by the nutria, go to the stomach, and then pass through the intestinal tract over the next few days. Flakes passed through some nutria quickly and could be detected in feces and intestines the same day the animal ate the bait. Flakes persisted in other nutria for at least 2 days. By using both flakes and tetracycline, the wildlife managers could mark the animals immediately with flakes and permanently with the tetracycline.[17] This type of information helps determine how far nutria travel after consuming the bait. Not all nutria die from the same levels of poison; some need more and others need less. In this study, 80% of the nutria that consumed the baits died. This surprised scientists, who expected that all of them would die.

The nutria in the study preferred sweet potato over apples and carrots. This was good news to the wildlife professionals, since the sweet potato was cheaper to use as bait and maintained a less soggy consistency than the carrots and apples which turned soft after a few days in the humid weather.

Nutria are social creatures that call to one another in their group. Scientists used recorded nutria calls to attract these rodents to the bait. The nutria either ignored or avoided the calls. Nutria may have calls specific to their group and may refuse to respond to calls from outsiders, a theory that would explain why the recorded calls were ineffective.

Nutria were attracted by smell. The nutria followed an odor trail of nutria, waste where the waste contained a distinctive odor recognizable to the nutria.

Partnerships to Manage Nutria Populations

Invasive animals such as the nutria do not honor the same boundaries that people do. Nutria do not know when they have crossed the border into Louisiana or when they have entered into federal land versus the property of a private landowner. Partnerships where agencies work together are the best way to combine resources for a better result.

The nutria cook-off was a project initiated by Louisiana's Coastal Restoration Program and funded by a law passed by Congress in 2000 called the Coastal Wetlands Planning, Protection, and Restoration Act. This law calls for state and federal partnerships. This project has three major components: (1) planning, developing, and conducting nutria meat marketing; (2) conducting a coast-wide survey to study habitat damage from nutria; and (3) providing financial incentives to trappers and nutria meat processors. The goal of the project is to create a market for nutria. People will then want to harvest it, and this will be one control mechanism to the mammal's explosive growth.

4 Round Goby

A FISH IN ROCKY WATERS

Melinda is a fisheries major at college and is just learning to dive. Today is her first day of diving in the Great Lakes. Prepared to see fish she has only learned about in textbooks, she shivers with excitement, waiting for the rest of the crew to gear up. After a quick count of three, she's in the water. Small fish swim away from her underwater intrusion—except one kind of fish. Melinda feels as if this frog-eyed creature is actually coming over to see what she is doing near the rocky bottom of the lake's edge. Melinda recognizes the invasive fish immediately: the round goby, an aggressive fish notorious for chasing other fish away from spawning grounds and habitats, as well as stealing bait from fishing hooks. The goby and Melinda stare back at each other and Melinda resists the urge to swim away, remembering that despite its brazen spirit, the goby presents no harm to a person, just to our aquatic ecosystems.

BIOLOGY AND BEHAVIOR OF THE ROUND GOBY

Round gobies (*Neogobius melanostomus*) are bottom-dwelling fish in marine and freshwater environments (Figure 4.1). The pelvic fins create suction, enabling gobies to attach to the bottom even in strong currents. The advantage of this suctioning ability is that they are able to feed in rougher currents.

Figure 4.1 The round goby is an aggressive, exotic species and is threatening native fish. The round goby suctions itself by the use of a fin to the bottom of a body of water to help it capture bottom-dwelling prey.

Their preferred underwater habitat is rock, sand, gravel, or crevices because they like to hide. They tend to stay near the shore, but during the winter they will migrate to deeper waters.

Young gobies are solid slate gray. As they age, they become mottled with black and brown. The dorsal fin is greenish with a black spot and no spines. The eyes are raised and resemble those of a frog.

The round goby is an aggressive fish that competes with native fish such as sculpins (*Cottus spp.*) and logperch (*Percina spp.*) for food. **Benthic**, or bottom-dwelling, organisms are the preferred diet of the adult goby. They will even prey on smaller gobies, although there is some indication that they are only able to detect prey when they are stationary. Juvenile gobies primarily consume benthic zooplankton and invertebrates.

Females reach sexual maturity at 1 to 2 years of age and males reach sexual maturity a little later, at 3 to 4 years of age. Gobies spawn every 20 days, and the males aggressively defend their nests.

BROUGHT BY BALLAST WATER

The round goby is a relatively recent invader from Eurasia (the continent that includes Europe and Asia), including the Black Sea and Caspian Sea. The first recorded collection was in the 1990s. According to the USGS, the round goby was first

A Natural Dilemma: When One Invasive Species Becomes the Predator of Another

Occasionally, an invasive species will eat another invasive species, leaving natural resource professionals wondering if this is a viable way to control the invasive prey. This is the case with the round goby, which happens to eat zebra mussels.

Zebra mussels are an approximately inch-long exotic mollusk that has already devastated native mussels in the Great Lakes. The zebra mussel is so abundant that beaches along the Great Lakes become covered with their sharp-edged shells and rotting masses of mussel flesh, forcing the beaches to be closed. Zebra mussels are also notorious for clogging water pipes because the mussels grow one on top of the other. Eventually the pipes become so clogged that water will not flow through them. The estimated damage within the Great Lakes region due to the Zebra mussel is in the billions of dollars.

In their native range, gobies consume zebra mussels as a considerable part of their diet. In lab experiments, a single goby will eat 78 zebra mussels in a day.*

introduced into the St. Clair River on the Michigan-Ontario border through ballast water.

The round goby is established and undergoing a population explosion in the Great Lakes. The environmental and economic impacts are already readily apparent. In places where the round goby is known to exist, there are marked declines in sculpin and logperch. Ohio has shut down the smallmouth bass fishery in Lake Erie during the months of May and June because of the goby's eating of smallmouth eggs. Under normal circumstances, male smallmouth bass guard nests and are effective in keeping

Even with this high consumption rate, gobies will not be enough to control the zebra mussels. In addition, zebra mussels are filter feeders and therefore consume toxins found in the Great Lakes. The zebra mussels accumulate the toxins. The gobies that eat a lot of zebra mussels will accumulate these toxins. The goby will have a higher concentration than the mussels because the toxins will magnify in their tissues. This is called biomagnification. The concern is that many fish that people eat, such as bass and trout, also eat the goby, and these fish would also have the toxins in their body. Consumption of these fish would equate to an intake of toxins from the Great Lakes.

Invasive species that seem to limit the numbers of other invasive species pose a dilemma to natural resource professionals. In the case of the round goby, this fish would still not be able to consume enough zebra mussels to be advantageous to any aquatic system.

***ANS Task Force. "Protect Your Waters." Available online at: http://www.protectyourwaters.net/hitchhikers/fish_round_goby.php.**

round gobies away. When males are removed, round gobies immediately invade and have been shown to eat up to 4,000 eggs within 15 minutes.[18] Because round gobies eat the eggs of lake trout, the survival of lake trout is further jeopardized in many areas, particularly the Great Lakes. The loss of trout in the Great Lakes would greatly hurt the region's economy.

Round Goby Management

Many natural resource agencies are enlisting the help of anglers to control the round goby population. Officials realize that the first step of controlling any invasive species is determining the distribution and formulating a strategy. Anglers can be "the eyes and ears" of fish and wildlife agencies that do not have the proper resources to inspect all bodies of water that might have gobies in it.

Alerts and fact sheets are sent to anglers asking them for help with looking out for the goby, along with a map showing the current distribution. The Minnesota Sea Grant Program provides round goby identification cards with a photo of the fish on the outside. When the card is opened, there is a detailed diagram of the goby. Regardless of the organization, the message is clear: If you catch a round goby outside of the areas on the map, preserve it in alcohol (grocery store alcohol or rubbing alcohol is fine) or freeze it. Then contact your state fisheries management agency or a Sea Grant institute. Check the Internet for the fisheries management agency in your state. Anglers should be prepared to describe when and where the goby was caught. Anglers are also prohibited from using the round goby as live bait.

Northern Snakehead

THE PREDATOR THAT TRAVELS OVER LAND

5

In 2002, a Maryland angler caught a long snakehead in a tiny retention pond behind a shopping mall. The experienced angler did not recognize the odd-looking fish and took some photos before releasing it back into the pond. Still curious a month later, the angler took the photos to the Maryland Department of Natural Resources (MDNR) headquarters in Annapolis for identification. One of the biologists feared it might be a snakehead. The photos were sent to the USGS in Florida, and it was confirmed that the fish was a northern snakehead (*Channa argus*).

Just a month after the first catch of northern snakehead, another angler was at the same retention pond and dipnetted eight juvenile northern snakeheads. The first confirmed report of both a northern snakehead and a reproducing population of snakeheads was declared within a month. Later that week, MDNR biologists captured 99 juveniles using electrofishing units. Biologists at both the state and federal level knew that they had to act quickly to prevent a widespread problem in Maryland waters.

MADE IN CHINA

The northern snakehead is native to the rivers of China southward and southwestward to the upper tributaries of the Yangtze River basin (Figure 5.1). It received its name due to

Figure 5.1 With a mouthful of teeth, the northern snakehead can eat fish up to one-third of its own body weight. One of the most amazing aspects of this fish is its ability to breathe air and wriggle across land.

its odd appearance, which includes a mouthful of teeth. This fish is also cultured in ponds, rice paddies, and reservoirs in China.

Although the northern snakehead is native to China, this fish is considered invasive in some of China's rivers. River ecosystems can vary widely within the same country. Plants and animals do not recognize the same boundaries as people. Chinese officials recognized the invasiveness of this fish in some of their waters in the 1920s and listed the fish as a careless importation. Today, reservoirs in China are full of snakeheads.

In 1997, in California's open waters of Spiritwood Lake, wildlife management employees captured a large snakehead using electroshocking. Since then, the northern snakehead has been reported in states all over the country. A specimen was found in Newton Pond, Massachusetts, after likely being released from a live-food market.

A Top-Level Predator

Juvenile snakeheads feed on small crustaceans and fish larvae; adults feed on fish, frogs, crustaceans, and aquatic insects. The northern snakehead can consume fish that are one-third of its own body length.

This species likes to feed in schools close to shore. These fish can breathe air and have been reported making a snorting sound as they rise to the water's edge to feed.

The northern snakehead is considered a voracious predator of fish and freshwater crustaceans. The biggest fear is that given this species' temperature and native range, it could potentially occupy all of the states in the contiguous United States.

Biological Functions

The most remarkable biological function of the northern snakehead is its ability to breathe air and wriggle slowly across land. Most fish are only able to access dissolved oxygen with gills. In fall 2005, a group of anglers fishing in a creek behind a trailer park saw numerous snakeheads and were shocked to see them coming up the bank. The anglers would catch one and throw it in the cooler and two more that were already in the cooler would jump out. They then had to chase them through the forest. This was one of the first known sightings in Virginia, although the fish was known to occur in neighboring District of Columbia.

The northern snakehead reaches sexual maturity between 2 and 3 years of age. This snake builds a mostly circular nest of aquatic plant pieces in shallow aquatic vegetation. Spawning occurs in the early morning; the female rises near the surface and releases eggs, which are then fertilized by the male. The average number of eggs released is roughly 7,000. In just 28 hours, the eggs hatch. Depending on the habitat, the northern snakehead has been recorded spawning anywhere from one to five times per season. Once the eggs hatch, the larvae stay in the nest and then leave the nest to feed on plankton. The larvae mature and develop the ability to breathe air. This occurs during what is considered the post-larval stage. Once this fish is able to breathe air, it begins feeding on small crustaceans and fish larvae.

Northern Snakehead Control

The most crucial step in controlling this fish is banning its import. This species continues to be imported for sale in live-fish markets. Fourteen states, including Texas, banned possession of live snakeheads prior to 2002. In the summer of 2001, game wardens from the Texas Parks and Wildlife Department discovered the northern snakehead in a Houston live-food fish market. The officials raided the markets and seized the fish as well as the shipping documentation. The distributor was traced and identified as an aquaculture facility that had imported northern snakeheads for sale for retailers in Texas, and agents confiscated those fish as well.[19]

Inspecting shipments into the United States is another way to control northern snakeheads. The following is one such inspection recorded by the USGS: In 2001, just before midnight in Blaine, Washington, a Canadian fish wholesaler declared a shipment of live ling cod (*Ophiodon elongatus*). The shipment was bound for Seattle. Michael Williams, a wildlife inspector for the U.S. Fish and Wildlife Service, inspected the shipment and found three open boxes that contained fish he thought were "unusual looking." When Williams asked the driver what the fish were, the driver's reply was that they were snakeheads that had been pond raised in China and shipped without water to Canada. Upon examining one box, Williams noticed that the fish moved and, on further investigation, found that most were alive and some "capable of vigorous movement." Williams informed the driver that possession of live snakeheads is in violation of Washington State regulations. The driver was asked to kill the fish and began striking them with a board. Williams notified the Washington Department of Fish and Wildlife and, after returning to the truck, found that the fish were still alive despite the driver's attempt to kill them. He seized the 80 fish and placed them in a freezer. When Washington Department of Fish and Wildlife authorities arrived, most of the fish in the freezer were still alive. State authorities took possession of the fish to proceed with penalties against the companies involved. The shipping invoice listed the fish as "Fresh Snakehead Fish – Product of China." The fish were subsequently identified as northern snakeheads.

Once established, northern snakeheads are difficult to control because they burrow in the mud at the bottom of a body of water. Pesticides provide a quick and relatively easy way to kill the fish. The dead fish float to the top, which helps verify effectiveness of the treatment. Unfortunately, native fish may also be killed.

Education is a crucial aspect of any invasive animal control program. The USGS investigates reports of northern snakehead in areas where it was previously not known to exist. The USGS partners with other wildlife agencies to implement a control strategy to minimize the damage the fish may have caused. The resources to control the fish are only as good as the educational program to alert anglers to the identity of the northern snakehead and why it is important to report it.

In southeastern Texas, an angler reported to the USGS that he thought he had caught a northern snakehead based on drawings he had seen in a book. He released the specimen without taking a photograph to confirm whether it truly was a northern snakehead. He also failed to follow up with a site location so that officials could examine the area. These incidents are disheartening to wildlife managers because allowing even one northern snakehead to exist can devastate an aquatic ecosystem.

Sea Lamprey

THE MARINE PARASITE WITH TEETH

6

The sea lamprey is an aquatic vertebrate that is native to the Atlantic Ocean. This parasite entered a tributary of the Great Lakes in the early 1800s. Lamprey influences on fisheries were not well documented in the Great Lakes until after the 1950s. As a result, sea lamprey populations became firmly established in the Great Lakes before anyone realized the harm they were doing to native fish. During the 1960s, observers on Lake Superior saw a 90% reduction in lake trout (*Salvelinus namaycush*) production while the number of sea lampreys caught was 70 times greater during that same time period. The decline was blamed on the sea lamprey, since the lake trout is a favorite host for the lamprey.

BIOLOGY AND BEHAVIOR OF THE SEA LAMPREY

Sea lampreys live in freshwater or salt water and go up rivers during the spring and early summer to spawn. Sexually mature lampreys seek stony, gravelly sites for mating. The lampreys carve out depressions in these soft soils, and the female lays up to 60,000 eggs. Adults die after spawning.

After the eggs hatch, the minute larvae live for several years as filter feeders near the soft sediments of the lake or tributary. As filter feeders, they feed on algae, dead or decaying organic matter, and various organisms that are in the stream water.

Larvae metamorphose into juveniles with eyes and teeth. Most important, a sucking disk develops. No more filter feeding for the juvenile lamprey as it begins to swim downstream. The lamprey now searches the waters for prey.

An adult sea lamprey feeds by attaching to other fish with its suctorial mouth. The mouth flares out when it attaches to a host with suction. Their teeth also create a wound kept open by anticoagulant generated by a gland. The lamprey extracts blood and other body fluids from the fish. Sometimes this parasitic relationship proves fatal and sometimes it does not. It depends on the site of the attachment on the fish's body, the time of year, and the size of the sea lamprey relative to the size of its prey. The sea lamprey's life cycle from egg to adult lasts from 5 to 8 years.

The Sea Lamprey and the Great Lakes

The sea lamprey is destructive to the Great Lakes ecosystem, in large part because of the injury and death it causes the lake trout. During its life, one sea lamprey can kill 40 or more pounds of fish. The lake trout is the preferred but not the only fish the lamprey will parasitize (Figure 6.1). As the trout's population declines, the lamprey parasitizes lake whitefish, chubs, and lake herring. These fish suffer declines and further offset the delicate aquatic food chain in the Great Lakes. Sea lampreys are so ruthless in their sucking that only one out of seven fish attacked will survive.

The decline of the lake trout further exacerbated another invasive fish problem: the predatory alewife (*Alosa pseudoharengus*). The alewife was native to the salty Atlantic and had spread throughout Lakes Huron and Michigan. The trout was controlling alewife numbers somewhat, but as the trout's numbers decreased, the alewife population exploded.

Figure 6.1 Lake trout is an important food source for both human and nonhuman animals. The lake trout also happens to be a favorite host of the sea lamprey. In waters where the two species coexist, a dramatic reduction in lake trout populations has been noted.

The alewives began to eat other native fish, and the native fish numbers declined significantly.

Sea Lamprey Management

In 1948, the Canadian and U.S. governments established a committee to resolve the lamprey problem. Research began on lamprey biology and specifically its spawning habits so that physical barriers could be put in place to prevent them from entering streams to lay eggs. The committee decided to use pesticides to kill the lamprey. They wanted something that would not be poisonous to other species and set out to research that possibility.

Sea lampreys have decimated trout populations in the Great Lakes and, as a result, the Great Lakes Fisheries Commission requested assistance from the Fishery Resources Office to control sea lampreys. Early efforts at sea lamprey control were concentrated on prevention of spawning by adult lampreys. Mechanical weirs were used to prevent spawning.

The weir was constructed of wire mesh on a V-shaped frame that directed fish and sea lampreys toward a trap. A wire mesh trap with a cone entrance allowed sea lampreys and fish to enter and prevented their escape. Spawning fish were transferred upstream of the weir, whereas the sea lampreys were destroyed. Debris would often clog the wire mesh of the weir, creating a dam-like effect that either collapsed the structure or caused flooding around it. The mechanical weirs were eventually replaced with electrical weirs.

Electrical weirs were constructed on large rivers to block spawning. Electrified rods are suspended in the water to direct fish and lampreys toward a trap or to stun them if they attempted to pass the weir. Just like with the mechanical weir, spawning fish and sea lampreys were captured in a wire mesh trap, and the sea lampreys were destroyed. The disadvantage of the electrical weir is that a loss of power allows the lampreys to pass around the weirs. Spawning fish would occasionally be killed or deformed by trying to pass through the electrical field. More sophisticated weirs are being evaluated by the U.S. Fish and Wildlife Service and Michigan.[20]

Lampricide was the next tool used to control the sea lamprey. In the late 1950s, more than 6,000 chemicals were tested, and one **selective toxicant** known as TFM was found to kill the lamprey. The lampricide works by killing the sea lamprey larvae that lives in the mud for 5 years. When a toxicant or poison is described as selective, it means that the pesticide kills a specific number of species, instead of all of them. In the case of a lampricide, a selective toxicant would kill only the lamprey instead of all of the fish. In the case of TFM, studies have shown that most larval sea lampreys are eliminated during treatments and most nontarget organisms survive.[21] Finding a pesticide to kill just one species is very difficult, particularly if the pest is biologically similar to other organisms that will come into contact with the pesticide. TFM has remained the primary method of sea lamprey control for about four decades.

Integrated pest management

Although lampricide is an effective way to control the sea lamprey, alternative methods of control are being sought to reduce pesticide use. The ideal situation is a combination of lampricides and other nonchemical methods of control.

The push to use fewer pesticides began in the 1970s and led to the concept of **integrated pest management** (IPM). IPM is a strategy to control plants, animals, and microorganisms by using chemical, mechanical, cultural, and biological methods. Chemical control methods are pesticides; mechanical methods include barriers in streams; cultural methods are strategies that are used on a repeated basis, such as flooding or drawdown to

Biological Control: Are the Risks Worth the Benefit?

The danger of biological control is that it requires introducing another nonnative species to solve the original problem of an exotic introduction. This new species can become a pest as well and potentially more difficult to control.

The classic "oops" biological control story involves the mongoose, rats, and Hawaii. The small Indian mongoose (*Herpestes auropunctatus*) is a slender animal with short legs and small rounded ears located on the side of the head. The mongoose is a solitary creature and has well-developed teeth used to tear flesh. This animal was introduced to Hawaii during the 1870s to control Norway rats in sugarcane. The mongoose ate the Norway rats but did not eat the tree rat because the mongoose could not climb trees. The Norway rat and tree rat were competitors and because the Norway rat was being eliminated, the tree rat was able to increase and became a pest in Hawaii. The mongoose also ate snakes and snake eggs, decreasing the snakes that were previously controlling the tree rats. As if all of that was not bad enough, the mongoose then became a pest in Puerto Rico, where it ate chickens and waterfowl and also became a vector for rabies. Much has been learned since the 1870s, however, and mistakes like the intentional introduction of the mongoose no longer happen.

The biggest concern is that the organism introduced to control the pest will become a problem itself. Luckily, scientists have experiments they can do to see if this might happen. Usually these experiments take up to 10 years to complete. Even then, there are no guarantees; nobody knows for sure how the organism will act in a real-world environment versus in a lab. Scientists generally recommend only introducing biological control agents that attack or eat only the target pest.

kill plants or animals; biological control introduces a natural predator to control the pest.

Low-head barriers are a mechanical control method for sea lamprey management. These barriers block spawning runs of adult sea lampreys while allowing passage of desirable fish such as the rainbow trout.

The Sterile Male Release program is another control technique implemented in the 1990s. Male sea lampreys are caught and transported in fish trucks to the Sterile Male Facility in Michigan, where they are injected with the sterilant bisazir. Sterilized males are released back into the water. Eggs are periodically sampled to determine if they are viable to evaluate the success of the program.

Another important part of any IPM program is to understand the current distribution of the pest. Mapping the distribution of sea lampreys is much more difficult than mapping weeds in a field, but it is possible. With the use of electrofishing gear, geographic information systems, and global positioning systems technologies, distributions of lampreys can be geographically referenced and downloaded for display on a map.

By using IPM strategies and improving the efficiency of pesticide treatments, TFM applications have been reduced by more than 25% since 1990.

7 Digging Deep

THE CHINESE MITTEN CRAB

It is 1992 in the San Francisco South Bay and a perfect day to net shrimp. The shrimp trawler is expecting a sizable catch of bay shrimp. Last year was a poor one for catching bay shrimp, but other trawlers told him that this year looked promising. Nobody really knows why the abundance of bay shrimp varies so much from year to year, although some of the old-timers claim it has to do with how much the river flows into the estuary. Supposedly, the bay shrimp thrive when the bay receives more freshwater from the river; the low salinity favors the rearing of juvenile bay shrimp.

The trawler prefers Redwood Creek, where waters are less than 20 feet deep in channels of the estuary's shallow reaches. The sunlight penetrates the clear, shallow water, and the bustle of city life feels far away. The water looks slightly murky in one area and he thinks he sees something burrowing in the sediment. He draws his net up to get his first catch of the season and sees a tiny, furry-clawed crab. He does not realize it, but it is the Chinese mitten crab (*Eriocheir sinensis*) (Figure 7.1). The darn thing got stuck in his net and did not drop out when he shook it. He plucks it out of the net and inspects it. Nothing he has seen before, but he's no biologist. The trawler shrugs and throws the fuzzy-clawed crab back into the water. This 1992 catch was not reported until many years later.

Figure 7.1 A notable feature of the Chinese mitten crab is the dense hair patches on its white-tipped claws. The juveniles' burrowing habits erode riverbanks in the San Francisco Bay.

It is now 1994, and that mitten crab from 1992 could have spawned hundreds of thousands of eggs that would become juvenile mitten crabs. This year, 1994, is a lucky one for the bay and perhaps an unlucky one for the mitten crab. Another shrimp trawler happens upon a Chinese mitten crab in his net and brings it to the California Academy of Science in San Francisco, where they properly identify it. This is the first reported occurrence of the Chinese mitten crab in the San Francisco Bay, although it is likely that the Chinese mitten crab was present even before the 1992 discovery, probably those who came across it did not know what they were looking at.

BIOLOGY AND BEHAVIOR OF THE CHINESE MITTEN CRAB

According to the World Conservation Union, the Chinese mitten crab is one of the 100 worst invasive species. Native to China and Korea, this crab is named for the dense patches of hair on its white-tipped claws. Although this crustacean may have a benign-sounding name, it causes significant damage to fisheries and native aquatic systems. It is established in the San Francisco Bay and Delta watershed, posing a threat to the estuarine communities and fishing operations.

The mitten crab is not a picky eater and has a diet that includes algae, dead organic matter, worms, and clams. Because it eats both plants and animals, the mitten crab is classified as an omnivore. In addition, the mitten crab is believed to eat fish eggs, specifically salmon and sturgeon eggs. Ecologists believe the mitten crab impacts freshwater and estuarine food webs at many levels and leads to the extinction of native invertebrates.

Although this crab is only 3 inches (7.5 centimeters) long, its high numbers and extensive burrowing cause erosion of levees and stream banks. The mitten crab interferes with water diversion projects in California. The crabs get caught in holding tanks meant to save the lives of fish from turbines. When the tanks get filled with the mitten crabs, the fish get killed in the turbines.

Like many invasive species, the mitten crab tolerates a wide range of living conditions but tends to become abundant in large estuary and river systems. The climate in all the known occurrences of the mitten crab is temperate, which increases the likelihood that it can live in many temperate aquatic systems in the United States. This crab also has a high tolerance for extremely polluted waters.

The mitten crab does have habitat preferences, however. Juvenile to adult-stage mitten crabs prefer slow-moving, warm, shallow waters up to 6.5 feet (2 meters) deep with submerged vegetation. Fast-flowing, cold-water rivers have been thought to be undesirable for rearing mitten crabs. Scientists have found in Asia, Europe, and North America that watersheds supporting large mitten crab populations have rivers that are more than 188 miles (300 kilometers) in length. In China, populations of crabs are found in most coastal streams and rivers but these populations are relatively small, and the crabs do not live as long nor grow as large as in the major population centers in the large river systems.

Adult crabs live in freshwater but migrate downstream to reproduce in brackish waters of estuaries. Animals that live in freshwater but breed in salty waters are described as **catadromous**. The salinity requirements reported for mitten crabs vary with their life stage. The highest salinity content is required for spawning. Spawning is the laying of the eggs and their attachment to hairs on small abdominal limbs of the mitten crab. Many scientists believe the salt enables the eggs to cement to the small abdominal limbs, known as pleopods, on the crab. Experiments with much lower salinity contents have shown that the eggs fall off the pleopods.

The females carry up to a million eggs until they hatch. The crab eggs hatch and the larvae spend their early life in brackish waters. It can take anywhere from 1 to 5 years to reach sexual maturity and then the mitten crabs go to brackish or salty waters to breed, sometimes traveling as far as 800 miles (1,330 kilometers). Both male and female crabs die soon after reproduction.

Mitten crabs are adept walkers and can get around most obstacles such as dams. In Germany, large numbers of mitten crabs left the water at night when they encountered an obstruction, and occasionally wandered the streets and entered houses. In California, mitten crabs have been found on roads and airport runways, in parking lots, yards, and swimming pools.[22] From a control standpoint, the one advantage of their ability to walk is that raccoons and river otters can then catch and eat them. Other animals that prey upon the mitten crab are sturgeon, striped bass, channel catfish, bullfrogs, and wading birds.

One of the reasons that the mitten crab is such a concern is due to its history in Germany. The first mitten crab collected in Europe was found in Germany in a tributary in 1912. Within the next couple of decades the population had exploded and spread to many northern European rivers and estuaries. Mitten crabs have become established in Europe at latitudes of up to 54°N, exceeding the northernmost distribution of crabs in their native range by 12°.

In Europe, mitten crabs get caught in fishing nets. In the San Francisco Bay, removing crabs from the nets has been time-consuming and costly to shrimp trawlers. One trawler reported catching more than 200 crabs several times in a single tow. The mitten crabs can be caught in such high numbers that they can harm or kill the catch.

Oriental Lung Fluke

The mitten crab is also the host for the Oriental lung fluke, which is a human health concern. Humans can become infected by eating raw or poorly cooked crabs. Lung flukes are parasites with a complex life cycle that requires them to infect two intermediate hosts before they mature into adults inside a person or other warm-blooded animal. Larval flukes infect snails, then move on to crustaceans, such as the mitten crab. In mammals, the breeding of the liver flukes destroys the respiratory system of the host.

In Asia, people enjoy mitten crabs as a delicacy. Even in the United States, live female mitten crabs with eggs can sell for $20 in Asian markets despite the fact that it is illegal to sell them.

Individuals who eat infected crabs suffer from lung damage. The flukes can even spread to the brain. Health officials became concerned that the health of recreational fishers, who eat the mitten crabs, could be in jeopardy.

Scientists analyzed hundreds of freshwater snails in the San Francisco Bay to find evidence of the lung flukes and found nothing. They also dissected crayfish and more than 800 mitten crabs and found no lung flukes.

The fact that the lung fluke has not made it to the San Francisco Bay is not surprising given what we know about invasive species. The lung fluke is a parasite of the mitten crab that helps keep the crab population limited. It is common for invasive species to be introduced without their predators. That is one of the key reasons they tend to thrive in their new environment. Identifying the predator of an invasive species can provide a biological control method but because of the health dangers of the lung fluke, this is one predator that will never be introduced.

Mitten Crab Control

Mitten crabs have been trapped as they migrate upstream. Legislative restrictions offer some hope for slowing the spread. In California, it is illegal to possess and transport a live mitten crab.

Ballast water contributes greatly to the introduction and spread of the Chinese mitten crab. By exchanging ballast water in the open seas, new introductions of mitten crabs in ports could be halted.

Chemical control has never been a realistic option because of how tiny the crabs are and how easily they disperse. Researching lethal fungi is under way and may offer a biological control method.

8 Managing Invasive Aquatic and Wetland Animals

Susan McGrath sees invasive species wherever she goes. Going for a run down a dirt path at her family's old New Hampshire camp, she could not just relax and enjoy the landscape. McGrath stopped because she saw a weed she had never seen before in that location. She pulled it because she remembered that scientists say early invasions can be halted if the land is purged of the invasive species—much easier to do with a plant than with an animal. For the rest of her vacation she could not go running without looking for that weed. If she spotted it, she hunkered down on the ground and pulled it out with all of her strength. She even put the ones that had gone to seed in her pocket.

McGrath's obsession, as some would call it, started when she began researching an article on invasive species for *National Geographic*. McGrath's heightened awareness is common among people in the field who deal with the problem of invasive species on a regular basis. They see invasive species everywhere they go. Unfortunately, this is not hard to do. Drive along almost any American roadside and the purple blossoms of Canada thistle are often as tall as the average woman. Travel to Jamaica and the same species of carp that is uprooting aquatic vegetation and hence dirtying the waters of the United States is thriving. Zebra mussels compete with zooplankton for food in the Great Lakes, as well as in the waters of Western Europe.

AWARENESS OF INVASIVE SPECIES

Awareness is a crucial part of any environmental campaign. Increasing public awareness of invasive species through education and outreach programs is a priority, according to the U.S. Fish and Wildlife Service. McGrath is a good example of how people can change their behavior based on their education about an issue. Based on McGrath's knowledge, she began pulling an invasive plant along a dirt road. The recreationist who moves his boat from one lake to another might take an extra moment to pull the weeds off the propeller. This simple task could prevent a new weed introduction in another body of water. The angler who spots what he thinks is a northern snakehead can save an aquatic ecosystem by early detection and notification to a state fisheries office. These seemingly simple actions by ordinary people can greatly help prevent the spread of invasive species. An informed public can help stop the spread of invasive species, and fewer overall resources will have to be used by agencies to control invaders.

Awareness of the issue of invasive species occurs through numerous media outlets: radio, television, newspapers, and magazines. Conservation organizations and natural resource governmental organizations often increase awareness with programs aimed to control invasive species in their region. Using that awareness to control invasive species is an ongoing task that multiplies its effect. One person who learns about invasive species may go on to share this information or the resulting behavior with another person. Putting effort into educating the public about invasive species has profound effects.

Workshops and meetings are another way to educate the public about invasive species. The Aquatic Nuisance Species (ANS) program within the U.S. Fish and Wildlife Service coordinates programs that target important issues such as State ANS plan development preventing the spread of aquatic nuisance species through the New York canal system and limiting mitten crab impact on the West Coast. ANS program coordinators also attend boating and sportfishing shows, informing thousands of participants about the definition, biology, and impacts of aquatic nuisance species and what they can do to help prevent their spread.

INVASIVE SPECIES DATABASE

The current distribution of invasive aquatic and wetlands species is still not well understood. The first step in creating an invasive species management plan is to get an idea of the distribution and extent of a species' population. The USGS created an information resource and database specifically for invasive aquatic species. The database was created from the passage of the Nonindigenous Aquatic Nuisance Species Control and Prevention Act of 1990. The Act created the Aquatic Nuisance Species Task Force to come up with creative solutions for tackling the problem of invasive aquatic species.

The Task Force believes a database that is both highly accurate in terms of location and timely are critical. Currently, the database contains locality information on more than 1,100 species introduced as early as 1850.

The best way to get a high degree of accuracy in terms of where exactly the location is on Earth is through global positioning systems (GPS) and geographic information systems (GIS) technologies. GPS refers to the hardware that is used when a person is outside collecting data. GPS equipment usually consists of a handheld device that collects electronic data by pinpointing the location using satellites. GPS data is then downloaded to a map showing the geographically referenced locations on a map. A program that is able to read the data that comes from a GPS unit is referred to as GIS, because the program can interpret geographically referenced data.

An example of what a person might have after downloading GPS data to a GIS program is a map with points reflecting the presence of invasive species.

Ideally, the task force would like to keep a file of each species with publications, correspondence, and any computer data collected. This information would be available to interested parties.

Data will be collected from researchers, field biologists, fishermen, and others involved in activities in the aquatic environment. Universities and state fish and wildlife agencies will serve as a means of distributing the database. Specialists will be available to properly identify species.

The success of the database obviously depends on the ability of the task force to get the word out to people who can contribute information. People "in the know" will be actively solicited through personal communication, announcements in professional publications and other media, at technical meetings, in pamphlets, and by other appropriate means. Published reports in journals and museums will also be reviewed.

Once the information is collected, natural resource professionals that are involved with resource management will be alerted to confirmed sightings of species and what the impacts are.

EARLY DETECTION AND RAPID RESPONSE

The state of Washington knew it had to react quickly when the exotic European green crab (*Carcinus maenas*) was discovered in its coastal waters. The early implementation of monitoring and control efforts in infested bays has kept populations down. When Canadians discovered the European green crab on Vancouver Island, they quickly contracted with Fisheries and Oceans Canada to assist with the development of a monitoring and control plan in shared waters of British Columbia. Volunteers trained by Washington Department of Fish and Wildlife staff monitor more than 100 sites in Puget Sound for green crabs. One of the tasks in the plan is rapid response if the green crab is discovered.

Agencies can prepare for the arrival of exotic species they think are likely to invade an area. If the invasive species is known to live nearby or can live in a similar environment, having a way to control the invasive animal before it arrives is a good plan. Washington state has prepared for the freshwater zebra mussel (*Dreissena polymorpha*) based on the fact that it is well suited for survival in Washington waters. As such, a volunteer monitoring program is in place along the Columbia and Snake Rivers and in several lakes throughout the state. In addition, state inspectors check commercially hauled boats at entry points into the state for zebra mussels. State agencies once found zebra mussels on a boat traveling through Washington to British Columbia. Washington and Canadian authorities stopped the boat and removed the mussels before it infested new waters.

Washington and Canada are partnering to halt the spread of zebra mussels, which is common in coordinating management efforts. Bringing agencies together combines resources to amplify control efforts. For example, in waters near Chicago, partners coordinate sampling efforts to track the distribution of round gobies. The 100th Meridian Initiative is a partnership created to control the westward spread of zebra mussels and other aquatic nuisance species by boats and personal aircraft. Boat inspections and surveys are conducted in the 100th Meridian jurisdictions, including North and South Dakota, Nebraska, Kansas, Oklahoma, and Texas. Through inspections and boaters' surveys, partners can determine whether the boat came from, for example, zebra mussel-infested waters.

BALLAST WATER LEGISLATION

One solution to the problem of introducing invasive species is for ships to discharge and refill their ballast water in the open ocean. The United States and the United Nations International Maritime Organization recommend this practice, known as open sea exchange; however, few countries

An Unlikely Invader: The Snowflake Coral

Any species can become a problem if put in the right (or wrong, depending on how you look at it) environment. Campaigns to save coral are widespread. Therefore, it may come as a surprise that there could be such a thing as bad coral.

The snowflake coral (*Carijoa riisei*) is native to the Atlantic Ocean from Florida to Brazil and has invaded Hawaii's waters. It grows in the waters of harbors where ships dump their ballast waters. This coral can tolerate a lack of sunlight and has made its way from harbors to shipwrecks and shaded crevices on the deeper reefs.

The ecological impacts of this coral are not well documented, but scientists believe that the coral crowds space for other invertebrates. The snowflake coral has tiny stinging cells in the tentacles that enable it to capture of zooplankton for food. This coral is named for its ability to reproduce from polyps splitting in two, settling to the bottom and developing into a polyp.*

*P.L. Colin and L. Arneson. *Tropical Marine Vertebrates*. Beverly Hills, CA: Coral Reef Press, 1995. Available online. URL: http://www2.bishopmuseum.org/HBS/invertguide/species/carijoa_riisei.htm.

have adopted this practice. Environmental problems that require international cooperation are some of the most challenging.

Many states have adopted guidelines to enforce ballast water treatment. In 2000, the Washington state legislature passed ballast water management and monitoring legislation for vessels entering Washington waters. This law requires ships to do open sea exchange and report it. Exceptions are made when the open sea exchange would be unsafe, such as when the seas are stormy or rough.[23]

Ballast Water Treatment

No single management technique has proven successful in killing or removing all organisms in ballast water. Selecting a treatment method depends on the structural integrity of the ship, the size and expense of method, the amount of potential damage to the environment, the safety of the crew, and ease for port authorities to monitor compliance.

One option is open sea exchange. In the above example, instead of the ship discharging in the Great Lakes, it would empty the ballast water in the open sea, the Atlantic Ocean in this case, and then fill the ballast tank with ocean water. This method works because coastal organisms are unlikely to be able to survive in the open sea. The option has the appeal of being easily monitored since a simple salinity test would be able to detect whether the ballast water is seawater or freshwater. Another limitation is that sediments and residual water are difficult to remove from the ballast tank.

Biocides can also be used to treat ballast water by killing organisms. One concern is for the health of the crew handling these chemicals. Another concern is the potential of corroding the ballast tank. Ballast water also can be heated to temperatures between 95° F to 113°F to kill larger organisms such as fish, but not microorganisms. No easy solution to this major pathway of biological invasions exists. Until recently, ballast water was not given a high priority. Research continues for techniques to treat ballast water without jeopardizing the safety of crew members.

Hull Fouling

Sea ballast water accidentally carries organisms inside the ship, whereas hull fouling carries organisms on the outside of the ship afer they attach to the hull. Organisms such as barnacles, mussels, sea squirts, sponges, and algae are able to attach to the hull and be transported long distances. Once they arrive in a new port, they can create new exotic populations by releasing their larvae or attaching to another structure in the port. Hull fouling is easily solved by building metal hulls and using antifouling paints to prevent organisms from attaching to the hull. Open sea exchange will not reduce hull fouling. As shipping increases, it becomes more vitally important to incorporate measures such as antifouling paint to prevent an exotic species explosion.

Onshore Treatment of Ballast Water

Onshore treatment removes ballast water from the ship when it enters the port or prior to being loaded into a ship as ballast. The water could be treated in facilities dedicated to ballast water or in facilities for treatment of wastewater. Water can also be stored and recycled for use as ballast by other ships.

Although it sounds expensive to treat water, it costs about the same as other treatment methods. Onshore treatment has the advantage of not putting crew members' lives at risk by exchanging water on the open sea or requiring them to have hazardous materials on board. Cheap initial treatments such as sedimentation can be used onshore rather than on the ship because of the availability of onshore space. Sedimentation may be capable of removing many resistant life stages such as cysts and spores as well as organic and inorganic suspended sediment, making subsequent treatment (ultraviolet light or biocides) cheaper and more efficient. Biocides use chemicals to kill living organisms.

A non-chemical approach is the use of (UV) ultraviolet light where free-floating microbes are killed. This method allows water to pass over a fluorescent light. The light penetrates the microbes enough to mutate the part of the organism that

The Environmental Protection Agency: Regulating Pesticides

The Environmental Protection Agency (EPA) regulates pesticides. Prior to the creation of the EPA, there were no restrictions on the use of chemical compounds to pests. In 1970, the White House and Congress worked together to establish the EPA in response to the growing public demand for cleaner water, air, and land. Part of that public outcry was the result of Rachel Carson's 1962 book *Silent Spring*. Carson explored the connection between the use of DDT and injury to our ecosystems. DDT is an insecticide used primarily to kill mosquitoes. Carson explained in her book how DDT was able to enter the food chain and accumulate in the fatty tissues of animals, including humans, and cause cancer and genetic mutation. Although some of the details of the point are still debated, it is undeniable that Carson created public awareness that nature is vulnerable to human intervention.

enables it to grow and reproduce. UV may not always kill an organism right away but it will prevent its spread because it will not be able to reproduce.

Pesticides

Pesticides are a common way to control invasive species. Pesticide is a general term that refers to a chemical that kills pests and includes insecticides, termiticides, herbicides, rodenticides, and lampricides, among others. The effects are immediate. Compared with other control options, pesticides are one of the least expensive methods of control. There are some disadvantages to pesticides, however.

One problem with pesticides is death or injury to non-target animals. This is a particular problem in aquatic systems, where a direct application—as would be done to plants or through the use of baits with terrestrial animals—cannot be done. Some aquatic pesticides such as lampricides do specifically target lampreys rather than all aquatic life.

Another limitation is due to biological resistance. Biological resistance is when a living organism becomes tolerant to a chemical to which it was previously susceptible. Organisms adapt to their environment; they are able to change genetically to be more competitive in their environment. Given this, it makes sense that creatures would aim to be resistant to these synthetic compounds created to kill them.

Biological resistance develops when just a few animals, particularly insects, become resistant to the chemical. The resistant animals initially may be small in number, but will be the ones whose genes are passed on through reproduction. The new population will have the genes for resistance. Population growth is exponential, and the newly resistant animals quickly replace the susceptible animals.

9 Putting It All Together

SAVING THE SAN FRANCISCO BAY FROM INVADERS

On a pier in San Francisco Bay, people huddle around an orange washtub with tweezers in hand. Others argue over Latin names and another scowls at a coiled rope. They find sea squirts from Asia and the Atlantic; sponges from the Atlantic, and mussels that probably came from the Mediterranean. These dedicated people are marine biologists brought together to find out just what is living in San Francisco's waters. What they are finding is typical of what is found growing on hard surfaces or on docks in the central part of the San Francisco Bay: organisms mostly from some other part of the world.

A survey a little farther north in the Bay, in fresher or brackish waters, will unveil different species. Many of these organisms are growing on the sides of docks or hanging from the bottoms of boats and buoys: long, dangling, silt-covered strands that consist of an Atlantic hydroid (an early life stage of a jellyfish), covered by an Atlantic bryozoan (a moss-like animal). Moving in more protected waters, docks and rocks are coated with reef-like calcareous tubes built by a worm from the Indian Ocean.[24] The mudflats in the marshes around the bay are covered with the Atlantic mud snail, which is the most common snail in the San Francisco Bay.

Andrew Cohen, a marine biologist at the San Francisco Estuary Institute, organized the marine biologists to come together for a week to survey 15 sites around the bay to see if any new species have arrived or old ones seemed to have left since the last time they checked.[25] Cohen and other scientists believe that the San Francisco Bay and Delta is the most invaded estuary in the world. The San Francisco Bay Delta is the largest estuarine system on the west coast of North and South America and has a watershed that encompasses 40% of the area of California.[26] Coastal estuaries may also be particularly susceptible to the introduction of new species because of their relative geologic youth and low species diversity (Figure 9.1).[27] In general, ecosystems with more **biological diversity** (also called biodiversity) are more resistant to domination by invasive species.

Figure 9.1 This tidal mill rests on a river estuary where the river is met by the tides, causing saltwater and freshwater to mix. Estuaries are considered young in terms of geologic time, and therefore they are less likely to be as biologically diverse as other aquatic systems. Scientists believe ecosystems with less biological diversity are more vulnerable to invasive species.

GETTING THERE

Three-quarters of the commercial ships that reach the San Francisco Estuary are from ports on the North American west coast, and the rest are mostly from ports on the Pacific Rim, such as those in New Zealand, Australia, and Fiji. Although the number of ships arriving in the estuary has been gradually declining, the size of the ships and the cargo they handle has been increasing. In the next decade or so, exports will double in the estuary, which will result in a doubling of ballast water as well. Ballast water from these ships has been the primary mechanism for invasive aquatic species introductions. In the case of wooden ships, worms such as the shipworm hitch a ride by boring into the hulls of ships. Sponges, seaweed, barnacles, and mussels were all introduced through ballast water in the San Francisco Estuary.

Some invasive aquatic animals were intentionally introduced for food (Figure 9.2). Although the Chinese mitten crab came from ballast water to a large extent, it is also believed to have been introduced through live markets, as it is considered a delicacy in Asia. Another culprit is a species of oyster that was introduced to the San Francisco Estuary for food from all over the world. Oysters from Japan and the Atlantic coast were reared in the shallow waters of the San Francisco Bay, which was fenced off to keep out oyster-eating predators such as leopard sharks. The oysters themselves were never able to establish outside cultivated areas but many tiny organisms attached to them have become a serious problem. Many have spread to other bays along the Pacific coast.

THE INVADERS

The Asian clam (*Potamocorbula amurensis*) and European green crab (*Carcinus maenas*), in addition to the Chinese mitten crab and zebra mussel, have altered the ecology of the San Francisco Bay and Delta.

The Asian clam, originally from Japan, China, and Korea, was first found in the Bay and Delta in the 1980s. The Asian clam

Figure 9.2 Importing food into fish and seafood markets can bring invasive animals into the United States. The Chinese mitten crab is considered a delicacy in Asia and was brought here for sale in live markets.

population is so large that it can collectively filter the volume of the northern section of the estuary once per day. As they filter, the clams remove the plankton that is the base of the estuary's food web. Other bottom-dwelling organisms then have nothing to eat. The introduction of the Asian clam into the San Francisco Bay has resulted in dramatic changes to these soft sediment communities. It is thought to be responsible for the collapse of some commercial fisheries in addition to the decline in the diversity and abundance of many benthic species in the area.[28]

The Asian clam is described as being extremely tolerant or having a wide niche. This means that the clam can live in many habitats from nearly freshwater to high-salinity areas. The clam also tolerates polluted waters. Control is particularly difficult in these situations. The clam can even tolerate oxygen deprivation. This control strategy was used in ballast tanks since, it is likely

Biological Diversity: Invader Resistance?

Visitors to the Galápagos Islands, 1,500 miles (2,414 kilometers) west of the coast of Ecuador, must walk slowly over a foam antimicrobial mat before their passports are verified and their shoes touch Galápagos soil. The antimicrobial mat is an attempt to kill exotic species, perhaps a microbe or two that could foster disease and decimate the blue-footed booby, a bird native to the islands. Does such ecological richness come with a heightened vulnerability to exotic species invasions and the damage they can cause? Some ecologists believe that islands are more vulnerable to invasive species because the native species have not been evolving as much as their counterparts in continental ecosystems. Also, because island ecosystems are smaller, they are usually less biologically rich than a large one. Greater diversity conveys what is known as biotic resistance, which helps preserve the integrity of an ecosystem over time.

Other scientists argue that a smaller ecosystem simply means that there are fewer individuals and that a major event such as a hurricane may be more likely to wipe out an entire species. The Galápagos Islands have seen plenty of visitors over the centuries despite its remote location. These seafaring visitors have brought many invasives, mostly animals, such as the smooth-billed ani, a black bird that was meant to remove ticks from cattle but ended up preying on baby lizards, nearly pushing them to extinction. Cats and rats have also jumped off early cargo ships and led to the extinction of bird species.

Exotic species invasions do seem to affect islands more. Hawaii is well known for significant problems with invasive species. It is difficult to know if it is an issue of a small ecosystem or the result of a high number of exotics simply being accidentally put there. If either scenario is true, then perhaps islands are more vulnerable. One thing is clear, however: No easy equation exists to determine an ecosystem's vulnerability to exotics.

that the clam makes its way to the Bay and Delta via ballast water. In the case of the Asian clam, the oxygen deprivation treatment employed in the ballast tanks was unsuccessful.

The European green crab is another mighty invader. It is native to Europe and Africa and was most likely introduced to the San Francisco Bay and Delta through bait shipments in 1989. This hearty eater is causing the decline of native crabs. The green crab eats organisms from more than 100 families and 14 animal groups.[29] This voracious predator is able to crush mussels, which presents a threat to mussel farms in the area.

Particularly disturbing is the green crab's rate of spread. Most non-natives expand their range by 50 miles in 40 years. The green crab has spread much farther beyond that to harbors in Washington. This unprecedented spread is likely due to months of northward currents from the El Niño pattern. El Niño refers to a periodic warming of the ocean surface off the coast of South America. This warm current moves water from the western Pacific Ocean toward the east. El Niño enables large numbers of crab larvae to be guided to Washington.

As mentioned, the mitten crab also is devastating the economy and ecosystem in San Francisco Bay. In 1998, nearly one million adult mitten crabs got stuck in a collection facility. The first problem was removing the crabs from the facility that was meant for the passage of fish. When these collection facilities get clogged, desirable fish are injured or killed.

WORKING TOWARD A STABLE ESTUARY

It may seem strange to describe an estuary or ecosystem as stable or unstable. We think of chairs and tables as stable or unstable. An unstable chair is much more likely to fall apart, perhaps when pressure is applied—like when a person sits in it. Ecosystems can also be evaluated in terms of their stability. It might not be as readily apparent as the wobbly leg of a chair, but if it is unstable, an ecosystem can cease to function in providing habitat and food for organisms.

A stable ecosystem is one in which the interactions of plants, animals, and environment follow the natural predator-prey cycles and species become better adapted to their environment through changes over time. When nature's processes get out of whack, species populations crash, setting off a domino effect of change for other species in the community. If an insect that is a food source for a predatory insect goes extinct, that predatory insect will need another insect species to eat or it too will vanish or will at least leave the estuary. If another insect exists to take the place of the other, additional disturbances in the ecosystem will be minimized. This is why a biologically diverse ecosystem tends to be more stable. When there are more species in the mix, there is usually a smaller chance that the removal of one species will have disastrous implications for the rest of the organisms in the ecosystem. The San Francisco Bay and Delta is relatively young in geologic time, and therefore the biological diversity is relatively low as compared with other ecosystems that have been around much longer. The Bay and Delta are also considered less stable because of the continual additions of invasive species and the current population of invaders.

Ballast water is a threat to the stability of San Francisco's estuary. The San Francisco Bay Regional Water Quality Control Board lists ballast water that has not been treated or exchanged on the open seas as a priority pollutant that impairs the health of the ecosystem. Open sea exchange still presents some danger, but treating the ballast water like wastewater once it arrives at the port does not present a threat to any crew member and protects the health of the ecosystem.

Outreach is critical to controlling any invasive species. The San Francisco Bay and Delta Bioinvasions Information, Coordination, and Outreach Project (ICOP) was created to provide resources and information to agencies and the general public. ICOP was created to handle all invasive species, not just aquatic and wetland species. Although many outreach efforts focus on educating the general public, it is also critical to increase accessibility of information to natural resource professionals (Figure 9.3). Often, one agency will collect data and neglect to share it with other agencies that could use it for their management projects. The 2-year ICOP developed information tracking to provide these reports to all agencies involved in the monitoring or control of invasive species.

Figure 9.3 Outreach efforts, such as posting informational signs in recreational areas, are an excellent way to increase awareness and promote action to control invasive species. Outreach involves any activity that informs people about the dangers of invasive species and provides tips on what they can do to help.

Children and the general public were also the focus of outreach efforts. The plan was to create an aquatic nuisance species school curriculum for middle and high schools that focused specifically on zebra mussel monitoring and prevention. The general public would be reached through fact sheets, meetings, and a Web site.

Mass media outlets such as television, video, and Internet list-servs will get the word out about invasive species in the Bay and Delta to other professionals and the general public. ICOP also created a monthly "rap sheet" that summarizes all major events and information regarding aquatic nuisance species during the month. Training programs are another outreach tool is training programs. ICOP will teach a program on monitoring, preventing, and controlling zebra mussel infestations to the

Soliciting the Public's Help: Strategies That Do and Do Not Work

Scientists have examined different techniques to get people to behave more "environmentally." Some techniques that are currently used are social pressure (for example, mandatory recycling), material disincentives (such as levying consumption-based taxes), and physical barriers to non-conserving behavior (for example, high-occupancy vehicle lanes on commuter routes).

These methods can make people feel guilty even if no one is monitoring them. However, they may also cause a mental state known as *psychological reactance*.* This is defined as the motivational state of a person whose sense of freedom has been restrained.** This situation produces a response by which people show an increased desire for a forbidden alternative or a decreased desire for what they feel forced to do. Research also shows that people will find creative ways to "misbehave." For example, there have been reports of people in mandatory recycling programs who have allegedly filled the outside water industry, utilities, and natural resource professionals. The list of outreach opportunities for this project goes on and on. The blueprint for any invasive species management plan incorporates: (1) survey and mapping to determine the extent and invasive species present; (2) legislative changes if appropriate, such as open sea exchange; (3) outreach and education to enlist the help and support of the public, as well as other natural resource professionals; (4) research to fine-tune better control methods; (5) implementation of control methods to rid the ecosystem of the invasives and prevent the introduction of new ones; and (6) follow-up to evaluate success. If the current plan does not look like it is reducing the spread of current and new invasive species, the strategy must be changed.

circumference of a trash bag with recyclables and then filled the inside with trash.[***]

Scientists believe that if you want people to behave in a way that helps the environment, it is best to make them feel like they are part of the problem and the solution, rather than making it seem like a chore. If people understand the threat to the environment and what they can do to help, they are much more likely to take steps to help, such as reporting a northern snakehead if they find one while fishing.

*R. DeYoung." Changing Behavior and Making It Stick: The Conceptualization and Management of Conservation Behavior." *Environment and Behavior* 25, no.4 (1993): p. 501.

**Ibid.

*** Ibid.

NOTES

1. Richard Ellis, *Aquagenesis.* New York: Penguin Putnam, 2001, p. 1.
2. Ibid., p. 17.
3. Louise I. Gerdes, *Endangered Oceans.* Farmington Hills, Mich.: Greenhaven Press, 2004, p. 20.
4. John Garvin, "The Beautiful Sea Monster," *Times of the Islands.* (Spring 2005), p. 86.
5. Ibid., p. 88.
6. Bernd Heinrich, *Winter World.* New York: HarperCollins Publishers, 2003, p. 172.
7. Richard Ellis. *Aquagenesis*, p. 131.
8. Ibid., p. 175.
9. Ibid., p. 249.
10. Terry McCarthy, "Wild About Nature," *Time.* (October 3, 2005): p. 61.
11. Donald O. Rosenberry. "Malformed Frogs in Minnesota: An Update." U.S. Geological Survey Fact Sheet 043-01. Available online at: http://pubs.usgs.gov/fs/fs-043-01.
12. Richard Ellis, *Aquagenesis.* 2001, p. 3.
13. Niles Eldredge, *Fossils: The Evolution and Extinction of Species*, Princeton, N.J.: Princeton University Press, 1991, p. 105–106.
14. J. R. Singleton, "Production and Utilization of Waterfowl Food Plants on the East Texas Gulf Coast," *Journal of Wildlife Management* 15, no.1 (1951): 46–56.
15. Joel K. Bourne, Jr, "Gone with the Water," *National Geographic.* (October 2004): p. 102.
16. Ibid., p. 92.
17. USDA/APHIS, Wildlife Research Center. "Developing Tools and Strategies to Reduce Mammalian Impacts on Forest Resources," Available online at: http://www.aphis.usda.gov/ws/nwrc/research/forest_resources/index.html.
18. Pam Fuller, Amy Benson, and Erynn Maynard. "*Neogobius melanostomus.*" USGS Nonindigenous Aquatic Species Database. Available online at: http://nas.er.usgs.gov/queries/FactSheet.asp?speciesID=713.
19. U.S. Geological Survey. "*Channa argus* Northern Snakehead." Available online at: http://fisc.er.usgs.gov/snakehead_circ_1251/html/channa_argus.html.
20. U.S. Fish and Wildlife Service. *Mechanical and Electrical Weirs.* 2005. Available online at: http://www.fws.gov/midwest/Marquette/weir.html.
21. U.S. Fish and Wildlife Service. "Sea Lamprey Management," 2005. Available online at http://www.fws.gov/midwest/Marquette.
22. Stephan Gollasch. "*Eriocheir sinensis.*" Invasive Species Specialist Group. 2005. Available online at: http://www.issg.org/database/species.

23. Washington Department of Fish and Wildlife. "Washington State Aquatic Nuisance Species Management Plan." October 2001. Available online at: http://www.wdfw.wa.gov/fish/nuisxsum.
24. Andrew N. Cohen. "Exotic Organisms, Proceedings of California's Emerging Environmental Challenges: A Workshop to Identify Future Issues for CAL/EPA." Office of Environmental Health Hazard Assessment. California Environmental Protection Agency, February 1999. Available online at: http://www.oehha.ca.gov/pdf/proceed.pdf.
25. Susan McGrath, "Attack of the Alien Invaders," *National Geographic.* (March 2005): pp. 92–112.
26. Jodi Cassell, "Bay-Delta Bioinvasions Information, Coordination, and Outreach Project (ICOP)." University of California Sea Grant Extension Program, 2002. Available online at: http://calwater.ca.gov/Programs/EcosystemRestoration/2002_Final_Proposals/184_compilation.pdf.
27. B. Herbold and P.B. Moyle. "The Ecology of the Sacramento-San Joaquin Delta: A Community Profile." Slidell, LA. U.S. Fish and Wildlife Service Biological Report No. 85 (7.22), 1989.
28. C. L. Hewitt, R.B.Martin, C. Sliwa, F.R. McEnnulty, N.E. Murphy, T. Jones, and S. Cooper, editors. "An Information System for Marine Introductions in Australia," National Introduced Marine Pest Information System (NIMPIS). Available online at: http://crimp.marine.csiro.au/nimpis.
29. Stephan Gollasch. Invasive Species Specialist Group Ecology of *Carcinus maenas*. Duesternbrooker, Germany: Institute for Marine Research, 2005. Available online at: http://www.issg.org/database/species.

GLOSSARY

Adaptations Small changes a species makes over time to better compete in its environment.

Animalia One of the five kingdoms. It includes organisms with a hollow body cavity, such as jellyfish; worms with segmented bodies; arthropods, which have exoskeletons and jointed legs, such as grasshoppers, lobsters, and spiders; and vertebrates such as humans and other mammals, birds, reptiles, amphibians, and fish.

Annelids Worms with segmented bodies.

Arthropods Animals that have exoskeletons and jointed legs, such as grasshoppers, lobsters and spiders.

Ballast water Water that ships take up to maintain stability during transport; this water usually contains exotic organisms.

Benthic Bottom-dwelling.

Biological control The method of introducing an invasive species' natural predator in order to reduce the population of the invasive species.

Biological diversity or **biodiversity** Richness of life forms in nature.

Biomagnification The exponential increase in chemical pollutants from the water through living organisms. Pollutants exist in the water and are concentrated in the living tissues of animals. The higher in the food chain the animal is, the greater the concentration of pollutants.

Bogs Freshwater wetlands, often formed in old glacial lakes, characterized by spongy peat deposits, evergreen trees and shrubs, and a floor covered by a thick carpet of sphagnum moss.

Brackish water An area where freshwater and salt water mix, as in an estuary. The salinity of brackish water is less than marine water but more than freshwater.

Catadromous Animals that live in freshwater but breed in salty waters.

Coelenterates Organisms with a hollow body cavity, such as jellyfish.

Chordates Organisms with a nerve chord and usually a backbone, such as humans and other mammals, birds, reptiles, amphibians, and fish.

Commensalism Relationship between two species in which one species benefits and the other is unaffected.

Crustacean Large class (Crustacea) of mostly aquatic animals, such as lobsters, that have exoskeletons, two pairs of antennae, and modified appendages on each segment.

Delta Fan-shaped deposit created at the mouth of a river because it widens and therefore slows.

Ecological niche The role of an animal in its environment: what it eats, what it does to avoid predators or catch prey, what resources it uses, and how it interacts with other species.

Eutrophication Process in which a lake, stream, pond, or any other body of water becomes high in nutrients, whereby the amount of dissolved oxygen available for aquatic organisms is decreased.

Fens Freshwater peat-forming wetlands covered mostly by grasses, sedges, reeds, and wildflowers.

Feral Previously domesticated animal living in the wild.

Filter feeders Animals that get their food by picking up tiny organic material from the water they are living in.

Fungi One of the five kingdoms. It includes mushrooms, molds, and mildews.

Generalist species Plants and animals that can live in a wide variety of environments.

Integrated pest management A strategy to control plants, animals, and microorganisms by using chemical, mechanical, cultural, and biological methods.

Invasive species Animals, plants, and microbes that are infiltrators in ecosystems beyond their historic range. Their invasion threatens native ecosystems and/or commercial, agricultural, or recreational, activities dependent on these ecosystems.

Invertebrates Animals without a backbone that either have a jointed exoskeleton or are soft-bodied.

Marshes Wetlands dominated by soft-stemmed vegetation.

Monera One of the five kingdoms. It includes bacteria and blue-green algae.

Mutualism Relationship between two organisms in which both benefit. Also called mutually beneficial.

Natural selection Theory developed by Charles Darwin that states that less-well-adapted species gradually disappear, and are replaced by those that are more fit.

Notochord A primitive rod, the precursor to the spinal column.

Parasitism Relationship between two organisms in which one organism benefits, the parasite, to the detriment of the host; the parasite generally does not kill its host.

Plantae One of the five kingdoms. It includes mosses, ferns, and flowering plants.

Protista One of the five kingdoms. It includes mostly unicellular organisms, such as algae and amoebas.

Selective toxicant Pesticide that kills a specific species, instead of all species. In the case of a lampricide, a selective toxicant would kill only the lamprey instead of all fish.

Specialist species Plants and animals that can live only in a narrow range of environments.

Veligers Baby conch.

Vertebrates Animals with backbones.

Watershed Area drained by a river.

BIBLIOGRAPHY

Aquatic Nuisance Species Task Force. *Protect Your Waters.* Available online at: http://www.protectyourwaters.net/hitchhikers/fish_round_goby.php.

Bauers, Sandy. "Invaders All Around Us." *Philadelphia Inquirer* October 3, 2004.

Bourne, Joel K., Jr. "Gone with the Water." *National Geographic* (October 2004): 102.

Burdick, Alan. "The Truth About Invasive Species." *Discover* (May 2005): 34–41.

Cassell, Jodi. "Bay-Delta Bioinvasions Information, Coordination, and Outreach Project (ICOP)." University of California Sea Grant Extension Program, 2002. Available online at http://calwater.ca.gov/Programs/EcosystemRestoration/2002_

Cohen, Andrew N. "Exotic Organisms, Proceedings of California's Emerging Environmental Challenges: A Workshop to Identify Future Issues for CAL/EPA." Office of Environmental Health Hazard Assessment, California Environmental Protection Agency, February 1999. Available online at http://www.oehha.ca.gov/pdf/proceed.pdf.

Cohen, Andrew N. and Anna Weinstein. "The Potential Distribution of Chinese Mitten Crabs (*Eriocheir sinensis*) in Selected Waters of the Western United States with the U.S. Bureau of Reclamation Facilities," Tracy Fish Collection Facilities Studies v(21). United States Department of the Interior, Bureau of Reclamation, Mid-Pacific Region, November 2001.

Colin, Patrick L. and Charles L. Arneson. *Tropical Pacific Invertebrates*. Beverly Hills, Calif.: Coral Reef Press, 1995.

DeYoung, R. "Changing Behavior and Making It Stick: The Conceptualization and Management of Conservation Behavior." *Environment and Behavior* 25, no. 4 (1993): 485–505.

Eldredge, Niles, Murray Alcosser, and Stephen Jay Gould. *Fossils: The Evolution and Extinction of Species*. Princeton, N.J.: Princeton University Press, 1991.

Ellis, Richard. *Aquagenesis*. New York: Penguin Putnam, 2001.

Forsyth, Adrian and Ken Miyata. *Tropical Nature*. New York: Charles Scribner's Sons, 1984.

Fritts, Thomas H. "Problems and Innovastions in Controlling Brown Treesnakes in Guam." *Aquatic Nuisance Species Digest* 4, no. 1. (November 2000): 1–4.

Fuller, Pam, Amy Benson, and Erynn Maynard. "*Neogobius melanostomus*." U.S. Geological Survery Nonindigenous Aquatic Species Database. Available online at: http://nas.er.usgs.gov/queries/FactSheet.asp?speciesID=713.

Garvin, John. "The Beautiful Sea Monster." *Times of the Islands*. (Spring 2005): 86–88.

Gerdes, Louise. *Endangered Oceans*. Farmington Hills, Mich.: Greenhaven Press, 2004.

Gollasch, Stephan. "*Eriocheir sinensis*." Invasive Species Specialist Group. Available online at: http://www.issg.org/database/species.

Heinrich, B. *Winter World*. New York: HarperCollins Publishers, 2003.

Herbold., B, and P.B. Moyle. *The Ecology of the Sacramento-San Joaquin Delta: A Community Profile*. Slidell, La.: U.S. Fish and Wildlife Service Biological Report No. 85(7.22), 1989.

Hewitt, C.L., R.B. Martin, C. Sliwa, F.R. McEnnulty, N.E. Murphy, T. Jones, and S. Cooper, editors. "An Information System for

Marine Introductions in Australia." National Introduced Marine Pest Information System (NIMPIS). Available online at: http://crimp.marine.csiro.au/nimpis.

McCarthy, Terry. "Wild About Nature." *Time* (October 3, 2005): 61.

McGrath, Susan. "Attack of the Alien Invaders." *National Geographic* (March 2005): 92–117.

McGrath, Susan. "On Assignment: Alien Invaders." National Geographic.com. Available online at: http://magma.nationalgeographic.com/ngm/0503/feature5/assignment1.html.

Rosenberry, Donald O. "Malformed Frogs in Minnesota: An Update." U.S. Geological Survey Fact Sheet 043-01. Available online at: http://www.pubs.water.usgs.gov/fs/fs-043-01.

USDA/APHIS, National Wildlife Research Center. "Developing Tools and Strategies to Reduce Mammalian Impacts on Forest Resources." Available online at: http://www.aphis.usda.gov/ws/nwrc/research/forest_resources/index.html.

U.S. Fish and Wildlife Service. "Mechanical and Electrical Weirs." 2005. Available online at: http://www.fws.gov/midwest/Marquette/weir.html.

U.S. Fish and Wildlife Service. "Sea Lamprey Management Program." 2005. Available online at: http://www.fws.gov/midwest/Marquette.

U.S. Geological Survey. "*Channa argus* Northern Snakehead." Available online at: http:// www.fisc.er.usgs.gov/snakehead_circ_1251/html/channa_argus.html.

Washington Department of Fish and Wildlife. "Washington State Aquatic Nuisance Species Management Plan." October 2001. Available online at: http://www.wdfw.wa.gov/fish/nuisxsum.

FURTHER READING

Baskin, Yvonee. *A Plague of Rats and Rubbervines: The Growing Threat of Species Invasions*. Covelo, Calif: Shearwater Books, 2003.

Cox, George W. *Alien Species and Evolution: The Evolutionary Ecology of Exotic Plants, Animals, Microbes, and Interacting Native Species*. Washington, D.C.: Island Press, 2004.

Forsyth, Adrian and Ken Miyata. *Tropical Nature*. New York: Charles Scribner's Sons, 1984.

Mooney, Harold A. and Richard J. Hubbs. *Invasive Species in a Changing World*. Washington, D.C.: Island Press, 2000.

Wilson, Edward O. *The Diversity of Life*. Cambridge, Mass.: Belknap Press of Harvard University, 1992.

_________. *The Future of Life*. New York: Vintage Publishing, 2003.

WEB SITES

Cornell University — Biological Control

http://www.nysaes.cornell.edu/ent/biocontrol/info/biocont.html

Eco-Pros Invasive Non-Native Species

http://www.eco-pros.com/invasive_non-native_species.htm

The Global Invasive Species Database

http://www.issg.org/database/welcome

Invasive Species in the Pacific, including Hawaii

http://www2.ctahr.hawaii.edu/adap2/hottopics/invasive_species.htm

The Nature Conservancy (TNC)—Invasive Species

http://www.nature.org/initiatives/invasivespecies

USDA Animal and Plant Health Inspection Service (APHIS)

http://www.aphis.usda.gov/ppq

USDA APHIS Pest Tracker

http://ceris.purdue.edu/napis/index.html

USDA National Invasive Species Information Center

http://www.invasivespeciesinfo.gov

The Wildlife Society

http://www.wildlife.org/policy/index.cfm?tname=positionstatements&statement=ps14

Wisconsin Council on Invasive Species

http://invasivespecies.wi.gov/awareness

PICTURE CREDITS

page:

8: Arnold T. Drooz, USDA Forest Service, www.forestryimages.org
11: Andrew J. Martinez/Photo Researchers, Inc.
14: Bill Lea/Dembinsky Photo Associates
17: Marilyn & Maris Kazmers/Dembinsky Photo Associates
19: Jim Roetzel/Dembinsky Photo Associates
20: Mark J. Thomas/Dembinsky Photo Associates
24: Marilyn & Maris Kazmers, Susan Blanchet/Dembinsky Photo Associates
27: Chris Schnepf, University of Idaho, www.forestryimages.org
29: Steve Allen/Photo Researchers, Inc.
34: E. R. Degginger/Dembinsky Photo Associates
37: Dominique Braud/Dembinsky Photo Associates
39: Gerald J. Lenhard, wwwforestryimages.org
40: Dr. Marli Miller/Visuals Unlimited
45: Eric Engbretson, U.S. Fish and Wildlife Service, www.forestryimages.org
50: U.S. Geological Survey Archives, U.S. Geological Survey, www.forestryimages.org
57: Bryan & Cherry Alexander/Photo Researchers, Inc.
63: Courtesy John Chapman and Sylvia B. Yamada, Zoology Department, Oregon State University
79: Martin Bond/Photo Researchers, Inc.
81: Brad Templeton
85: John and Karen Hollingsworth, U.S. Fish and Wildlife Service

Cover: Yva Momatiuk & John Eastcott / Photo Researchers, Inc.

INDEX

ABOUT THE AUTHOR

Suellen May writes for agricultural and environmental publications. She is a graduate of the University of Vermont (B.S.) and Colorado State University (M.S.). She has worked in the environmental field for 15 years, including invasive species management for Larimer Country Open Lands in Colorado. She served as the Education Committee chairperson for the Colorado Weed Management Association. While living in Fort Collins, Colorado, she founded the Old Town Writer's Group. She lives with her son, Nate, in Bucks County, Pennsylvania. Readers can reach her at suellen0829@yahoo.com.